Innovation as Strategic Reflexivity

How important is strategic reflexivity to business development?

This book presents a new view of innovation, seeking to disclose how strategic reflexivity is embodied in specific innovation practices and management roles.

From an evolutionary point of view, the contributors argue that firms and organisations are increasingly forced to take into account the growing complexity of the environment. To do this, they must create strategies that interpret external expectations, but also deal with the internal reflexivity processes caused by innovation. The way to bridge between strategy and innovation, they suggest, is through strategic reflexivity.

The contributions, both theoretically and empirically based, range across a number of disciplines, including economics, business administration, innovation studies, management theory, sociology and political science. These are all united by a theoretical core: the perception that strategic reflexivity is vital to business development, and that innovation should be much more thoroughly analysed.

All those with a broad interest in innovation theory and technical change, including social scientists, policy makers and managers, will find this book valuable and thought-provoking reading.

Jon Sundbo is Professor of Business Administration and Innovation at Roskilde University. His special focus is on management, strategy and organisation. He researches in the fields of innovation, service management and development, and has written articles about these topics in major international journals. He has published two books within these fields, *The Organization of Innovation in Services* and *The Theory of Innovation Theory*.

Lars Fuglsang is Associate Professor in Political Science and Innovation at Roskilde University. His main research interest is the impact of technology on services, particularly public services, and how institutions are created to deal with the impact of technology on society. He has written books and papers in the field of science, technology and public policy, service development, and technology and regional integration.

Routledge Advances in Management and Business Studies

Innovation as Strategic Reflexivity

Edited by
Jon Sundbo and
Lars Fuglsang

London and New York

First published 2002
by Routledge
2 Park Square, Milton Park, Abingdon, Oxon, OX14 4RN

Simultaneously published in the USA and Canada
by Routledge
270 Madison Ave, New York NY 10016

Routledge is an imprint of the Taylor & Francis Group

Transferred to Digital Printing 2009

© 2002 Selection and editorial material, Jon Sundbo and
Lars Fuglsang; individual chapters, the contributors

Typeset in Baskerville by
Florence Production Ltd, Stoodleigh, Devon

British Library Cataloguing in Publication Data
A catalogue record for this book is available from the British Library

Library of Congress Cataloging in Publication Data
Innovation as strategic reflexivity/edited by Jon Sundbo and
Lars Fuglsang.
    p. cm.
Includes bibliographical references and index.
    1. Technological innovations–Management.   2. Creative ability
in business.   I. Sundbo, Jon.   II. Fuglsang, Lars.
HD45 .I53722 2002
658.4′063–dc21   2001048498

ISBN 0–415–27380–3

Contents

Figures

Tables

Notes on contributors

John Bessant is Professor of Technology Management at the University of Brighton in the UK. Originally a chemical engineer, he has been active in the field of research and consultancy in technology and innovation management for over 25 years. His areas of special interest include human resource and organisational development for innovation, inter-firm learning and enabling the emergence of agile enterprises. He is the author of 12 books – the latest is *Managing Innovation*, with J. Tidd and K. Pavitt (Wiley, 2001) – and many articles on the above topics, and has lectured and consulted widely around the world.

Susana Borrás is Associate Professor in Political Science at the Department of Social Sciences, Roskilde University, Denmark. Her current research interests are centred on EU policy-making and governance patterns, together with the political economy of the EU's technology policy. Recent and forthcoming literature includes: *The Globalising Learning Economy: Implications for Innovation Policy Brussels (Commission of the EU)*, with B.-Å. Lundvall (1998) (http://www.cordis.lu/tser/src/globec.htm); *Innovation Policies in Europe and the US: The New Agenda*, with P. Biegelbauer (eds) (forthcoming (2002) in Ashgate); and *The Innovation Policy of the EU* (forthcoming (2002) in Edward Elgar).

Bo Edvardsson is Professor of Business Administration. He holds a PhD in business management from Uppsala University, Sweden and is Research Director of the Service Research Center – CTF, Karlstad University, Sweden. He has published a number of research reports and is author/co-author of 13 books, for instance *Nordic School of Quality Management* (Lund: Studentlitteratur, 1999); and *New Service Development and Innovation in the New Economy* (Lund: Studentlitteratur, 2000). He has published a number of articles in scientific business journals, such as the *International Journal of Service Industry Management*, *R&D Management* and the *Service Industries Journal*. His current research interest lies in the field of service quality with the emphasis on service design and the development of new services.

Lars Fuglsang is Associate Professor in Political Science and Innovation at Roskilde University, Denmark. He has participated in several research projects on innovation and technology development and has written international articles on the topic; he has also participated in projects on service development, organisation and innovation. He has published articles in international journals on innovation, entrepreneurship and service management, particularly in the public sector, and is co-author of several books on innovation, organisation and technology in Danish.

Andrea Gallina is Research Fellow at the Department of Social Sciences, Roskilde University, Denmark. His research activities are concerned with innovation in industry and small and medium-sized enterprises' development; as well as on international economics and regional integration agreements. He is co-author of a book on southern European and Mediterranean industrial production systems, and has published a number of articles on the economics of SMEs in the Mediterranean countries.

Faïz Gallouj is Associate Professor of Economics at the University of Science and Technology of Lille, France. His main fields of research are R&D, innovation and performance in the service sector. He has been published in several publications, including *Research Policy, Science and Public Policy, Service Industries Journal, International Journal of Service Industry Management, European Journal of Innovation Management, Journal of Socioeconomics, Revue française d'économie*. He is also the author or co-author of several books on innovation in services.

Anders Gustafsson is Associate Professor of Business Economics in the Service Research Center at the University of Karlstad, Sweden. He holds a PhD in quality management from Linköping University, Sweden. He has published a number of research reports and is author/co-author of seven books, for instance *Conjoint Measurement* (Springer, 2001), *Improving Customer Satisfaction, Loyalty, and Profit* (Jossey-Bass, 2000), *Nordic School of Quality Management* (Lund: Studentlitteratur, 1999) and *New Service Development and Innovation in the New Economy* (Lund: Studentlitteratur, 2000). His current research topics are on the managerial usefulness of quality and customer satisfaction research, and competing through service development: customer-driven innovation.

Birgit Jæger is Associate Professor at the Department of Social Sciences, Roskilde University, Denmark. She is also affiliated to the Center for Local Institutional Research as a member of a research group working with technology and innovation. She has primarily been doing research in the field of social studies of technology, and is currently working on a research project founded by the Danish Research Council, concerning older people's use of ICT. She has recently published 'Some lessons of social experiments with technology', in L. Keeble and B. Loader (eds) *Community Informatics – Shaping Computer-Mediated Social Relations* (London: Routledge, 2001).

Sten Liljedahl received his PhD from the Institute of Economics, SLU, Uppsala, in 1994 and has since then been Head of Environmental Management at the County Council of Halmstad. His main research interest is environmental management.

Marius T.H. Meeus is Professor of Innovation and Organisation at the Department of Innovation Studies, Faculty of Geography, Utrecht University, The Netherlands. His research focuses on the development and empirical exploration of organisation theory applied to the innovative behaviour of firms. His most recent work includes articles and book chapters on patterns of interaction in regional innovation systems, theory formation in innovation sciences, the selection-adaptation debate and innovative performance, learning and proximity. He published in *Organization Studies, Technology Analysis and Strategic Management, Research Policy, Journal of Economic and Social Geography*, and *Papers in Regional Science*.

Harry Nyström is Professor of Marketing and Entrepreneurship at Karlstad University; and former Professor at the Institute of Economics, SLU, Uppsala. He has previously held professorships at Uppsala University and Oslo School of Business and received a doctorate from the Stockholm School of Economics in 1970. His main publications are *Creativity and Innovation* (Wiley, 1979) and *Technological and Market Innovation-Strategies for Product and Company Development* (Wiley, 1990). He has also published articles in major journals, such as *Journal of Marketing Research, R&D Management* and *Journal of Product Innovation Management*. His main research interest is in innovation management, company creativity and strategies for product and company development.

Leon Oerlemans is Associate Professor in Economics of Technology and Innovation at the Department of Innovation Studies, Utrecht University, The Netherlands. His major fields of research are the economic organisation of innovation, the relationships between spatial proximity and innovation, and issues related to technology and innovation policy. Some of his publications include 'On the spatial embeddedness of innovation networks. An exploration of the proximity effect' (co-authored by M. Meeus and F. Boekema), in *Journal of Economic and Social Geography*, 2001, 92(1); 'Patterns of interactive learning in a high tech region. An empirical exploration of an extended resource-based model' (co-authored by M. Meeus and J. Hage), in *Organisation Studies*, 2001, 22(1).

Bodil Sandén is Doctoral Candidate in Business Administration at the Service Research Center – CTF, Karlstad University, Sweden. Her thesis highlights strategic and tactical aspects of user involvement in service innovation.

Jon Sundbo is Professor of Business Administration and Innovation at Roskilde University. He has dealt with innovation and service business as research fields, and is Director of the Centre of Service Studies and coordinator of the research group on innovation and technology development in the Department of Social Sciences, Roskilde University. He has authored several international articles and three books on innovation in English, *The Theory of Innovation* (Edward Elgar, 1998), *Organisation of Innovation in Services* (Roskilde University Press, 1998) and *Strategic Management of Innovation* (Edward Elgar, 2001). He has also published articles on service firms, e.g. the standardisation-customisation issue, empowerment of employees in service firms, etc., in major international journals.

Paul Trott is Senior Lecturer at the Business School, University of Portsmouth. His PhD from Cranfield University received the ICI prize for 'PhD of the year'. He has worked in research and development (R&D) for ICI and Unilever. He is the author of many reports and publications in the area of innovation management and his book *Managing Innovation and New Product Development*, published by 'Financial Times Management', is now in its second edition.

1 Innovation as strategic reflexivity

Jon Sundbo and Lars Fuglsang

The aim of this book

In the first part we briefly introduce the theoretical and analytical concepts to be used in this book and we present an overview of the chapters. In Parts II and III the theoretical ideas are developed in greater depth.

The theoretical idea

The book presents a new perspective on innovation, a Schumpeter Mark III approach (cf. the discussion of Schumpeter I and II in Phillips 1971). We call this approach 'strategic reflexivity'. It sees innovation as a strategic response to change-processes in the market and society (characterised by uncertainties).

External changes increasingly become crucial for the firm's possibilities for growth and survival. Market and social developments are particularly important. These include new patterns of behaviour and ethical norms, political regulations and new knowledge. As markets become more mature and complex, and consumers more demanding and critical, consumers do not just accept innovations. Innovations have to be negotiated. Firms cannot merely adapt to markets as stable environments, they need to be more reflexive and strategic. By reflexive we mean that firms must take into consideration the changing manifestations of their actions on the market and in society. By strategic we mean that firms must consciously negotiate their role in the division of labour and in relation to customers.

This approach, we argue, is relevant to an understanding of innovation in the contemporary economy, which is characterised by knowledge, services and flexibility. We think much more attention could be devoted to strategy as a force of innovation under conditions of reflexivity. We do not mean to say, however, that strategic reflexivity as a theoretical concept is relevant to all forms of innovation and industrial development, particularly not when innovation is characterised by clear technological opportunities for the firm.

Strategic reflexivity is, we argue, already implicitly described by recent innovation theories, especially those which are market-oriented (e.g. Kotler 1983, Nyström 1990, Porter 1990, Sundbo 1998a). We want to make this approach more explicit. For that purpose, we emphasise economic, political and management factors and the approach has particularly benefited from sociological inspiration.

The sub-components of the approach, which will be developed further in this and other chapters of the book, are interaction, roles, complementarity and innovation management. Briefly, innovation can be seen as being inspired by interaction among relevant actors, i.e. people and firms that have certain roles and expectations about the future. This interaction is not to be understood as a smooth social interaction process inside a well-ordered, rule-governed system or organisation, but as a reflexive and often fragmented and conflict-based game. In order to ensure a role for themselves, actors formulate and negotiate strategies of innovation that relate to the expectations of others. Reflexivity is strategic when it involves systematic negotiations of actors' roles and expectations about these roles.

Strategic reflexivity is often formative, not purely reactive. Strategic reflexivity deals with the impressions a firm makes on the market and society, but it also seeks to manipulate and control these impressions.

A driving force behind strategic reflexivity is the attempt to ensure and create a role for the firm and the employees in the division of labour that exists in the innovation process. Strategic reflexivity is, in itself, a process that makes it possible for people to adjust and change their roles.

Roles have to fit in with each other and complementarity between them is critical, particularly in the service sector. Furthermore, economic actors have strategic interests in their given roles that may constitute a hallmark for them. To change a role therefore is risky for both the firm and the employee. Radical innovations are therefore rare, and if entrepreneurship exists it is often of the Kirzner-type, making up for insufficiencies and mistakes rather than creating entirely new structures (cf. Kirzner 1973).

Management is seen as a relevant activity from which strategic reflexivity is constructed. Innovation management interprets market opportunities, and develops strategies from these interpretations. Furthermore, innovation managers can 'direct' personnel through changes, helping them to find their 'roles', much as theatre or movie directors direct actors.

Innovation is a comprehensive concept. It means new products, new production and delivery processes, new organisational structures, new market behaviour and new materials, following Schumpeter's original definition (Schumpeter 1934).

Furthermore, manufacturing is no longer the dominant industrial activity. New activities, which include new forms of innovation such as in services and in the information and communication technology (ICT) sector, are appearing (as demonstrated in Chapters 9 and 14 by Gustafsson, Edvardsson and Sandén and Jæger). The public sector is also changing its activities and is becoming more innovative and development oriented in order to preserve a role for itself in the division of labour (as Fuglsang, Chapter 10, demonstrates).

During the 1980s and 1990s, innovation was almost exclusively understood as technological innovation (cf. Dosi *et al.* 1988). However, many important innovations are not technological innovations in the strict sense of the word. They are sometimes social, which means that they are constructed as a result of changes in social behaviour rather than by engineers or scientists (cf. Davenport 1993, Sundbo 1998b). Innovation is not exclusively something engineers and scientists do. It also includes improvements at the shop floor level, in the office, among service workers, in marketing etc., and is not closely connected to one dominant technological trajectory.

We intend to focus on the structure of that process, not the single steps in it. We argue that the structure of innovation has changed in the flexible service and knowledge-based economy from hierarchy to interaction and from planning and forecasting of technology to strategic reflexivity and foresight.

The structure and content of the book

In the chapters of this book, strategic reflexivity is analysed from several angles. Each chapter is intended to provide illustrative examples of the concept: At the level of the firm, we will see how firms organise and manage their innovation process and develop new roles to become more interactive. At the inter-firm level, we will examine firms' motivation for interaction. At the policy-level, we will see how the EU has attempted to change science and technology policy – from top-down science-oriented approaches to interactive innovation-oriented approaches. Together, the different chapters of the book should provide elements to an understanding of the concept strategic reflexivity at the firm level as well as the meso- and macro-levels.

In order to illustrate what we mean by innovation as strategic reflexivity, we have collected papers from different social science disciplines including economics, business administration, innovation studies, management theory, sociology and political science. The single chapters are *not* to be seen as elements of one consistent common theory. For one thing, each chapter rests on its own analytical tools and methodologies and should be judged against its particular standards. What the different chapters have in common, however, is a theoretical core, the perception that strategic reflexivity is important to business development and innovation and should be much more thoroughly analysed. Some of the chapters are purely theoretical; some are more directly based on empirical evidence. All of them have a theoretical ambition that relates to this core.

The book is divided into four parts. Part I is theoretical. In Chapter 2 Nyström discusses how contemporary innovation processes demand creative management. The challenges to innovation theory are seen from a post-modern perspective. Gallouj presents (in Chapter 3) a theoretical concept of the innovation process as interactive. His chapter focuses on the contribution of services to innovation, particularly the role of consultancy in innovating firms, and what the consequences are for innovation theory. In Chapter 4, Sundbo examines the nature of internal interaction. The concepts of strategy and dual structure are investigated as important elements of innovation.

Part II deals with the external oriented assets of the firm – those that attempt to catch and exploit the possibilities on the market. Nyström and Liljedahl show in Chapter 5 how reflexivity and strategy-making interconnect in the development of new products by the use of open strategies. In Chapter 6, Gallina investigates the special challenges of the external interaction that small and medium sised firms meet in the globalisation process. Trott discusses, in Chapter 7, how firms must be more strategically reflexive in their use of marketing research which in some situations constitutes an obstacle to innovation.

Part III deals with the internal interaction activities of the firm that constitute the innovation potential. Bessant discusses in Chapter 8 whether innovation can be managed. He emphasises routines as a concept to understand innovation processes. Gustafsson, Edvardsson and Sandén show how a service firm develops innovations through systematic interaction between employees and customers (Chapter 9). In Chapter 10 Fuglsang analyses how a public organisation, which provides home-help to the elderly, develops new roles to create a more innovative organisation. Meeus and Oerlemans demonstrate, in Chapter 11, that internal stakeholders play a role in the innovation process and analyse how the innovation process depends on their involvement. Mattsson continues along that line by analysing how innovation also implies internal conflicts in the organisation (Chapter 12).

Policy is treated in Part IV. Borras (Chapter 13) discusses how technology policy has changed from a science-oriented top-down approach to a more innovation-oriented, interactive bottom-up approach. From this new perspective, policy should create some conditions for firms that must find the right way to act, but strategic reflexivity and ideas are important to the actors. Finally, Jæger presents (Chapter 14) an analysis of a social IT experiment which demonstrates the role of strategy-making under reflexive conditions for technological innovation and diffusion of technology in a community.

The book is the result of a workshop at Roskilde University in Denmark in November 1999. The papers presented at the workshop have been changed and edited to develop a more coherent framework; a few others have been added.

Strategic reflexivity as an explanation of innovation

We present, here, a more sophisticated argument for this theoretical approach to innovation and relate it to earlier explanations.

Current challenges to innovation theory

For several years, science and technology have been seen as factors that can explain economic growth (or correlate with it). However, innovation, as a somewhat broader concept, has increasingly been stressed and is becoming the central category for explaining economic growth – even if innovation is more difficult to measure and deal with theoretically. Thus, we are urgently in need of an explanation of what innovation is, how it takes place, with what other factors it co-exists and what it responds to. Furthermore, theories of science and technology need to be re-examined as theories of innovation.

Most theories of innovation refer back to Schumpeter's explanation of economic development published in 1934. In Schumpeter's original ex-planation, entrepreneurship was seen as the motor of innovation. Later explanations have, however, argued that the pre-conditions of innovation have changed since Schumpeter described the foundations of capitalism in the nineteenth century. Innovation has become more complex and dependent on organisational structures rather than individual entrepre-neurs. During, between, and after the two world wars, innovation became increasingly organised in laboratories, R&D departments and institutions of science and technology (cf. e.g. Jewkes *et al.* 1969, Rosenberg 1976). Theories which examine these more complex institutional settings have therefore been introduced (e.g. Lundvall 1988, Nelson 1993). They tend to shift the focus from innovation (the Schumpeterian approach) to science and then to technology. At the micro-level, the role of the entrepreneur has been replaced by theories of the firm and the firm organisation. Resource-based theories of the firm have, for example, been discussed in order to explain innovation at the firm level (e.g. Teece and Pisano 1994). Learning theory, based on psychological and sociological learning theory and knowledge theory, has been among the latest approaches (e.g. Nonaka and Takeuchi 1995). They too call into question the individualistic explan-ation of innovation and provide tools to describe innovation as a more complex interactional process.

Still, many of these theoretical and empirical contributions often empha-sise narrow categories such as science, technology or firm resources. They also appear, implicitly, to describe innovation as a rational process. Firms act much in the same way or have just a few variants. There is, indeed, a strong influence from econometrics and correlation analysis that can only deal with a limited number of variables. Other approaches seek, more heroically, to maintain a more complex picture of innovation. Nevertheless, most of them tend, in our opinion, to end up with rather incomplete analytical categories – unsatisfactorily explained. Theories of networks, interaction frameworks or innovation systems all deal with important concepts, but they are not well accounted for. The reason may be that the authors of these theories mean to import these concepts into economic theory. Here the fight concerns equilibrium theory. Details about the concepts would therefore, in a sense, miss the point. It is obvious, however, that more fundamental explanations of these and other related concepts are possible by involving other areas of research, such as sociology.

Innovation theory has been incorporated into an evolutionary frame-work (cf. Andersen 1994, Metcalfe 1998) which is inspired by Schumpeter's work. Within that framework there has also been a search for one or a few factors that can explain innovation – the innovation 'genes' for example (Nelson and Winter 1982 is the most well-known). These attempts have provided useful, new, knowledge that nevertheless tends to be limited in terms of a theory of innovation as strategic reflexivity. In the evolutionary framework, innovation is governed by selection mechanisms. Only those actors that 'fit' the structure of that mechanism at a given moment in time are supposed to survive. Thus, for example, if Henry Ford had not invented the Ford T, other economic actors that fitted this scheme (cheap stan-dardised cars) would have done so, since, in retrospect, this was the only possible survival strategy in this period. Consequently, it seems irrelevant for other than purely historical reasons to examine any alternative strategy at the micro-level.

Research has, however, convincingly argued that choices at the macro-, meso- and micro-level are possible (cf. Piore and Sabel 1984, Hirst and Zeitlin 1988, Penrose 1959), including choices that deviate from the main stream of innovation. A 'selection environment' does not exist in any definite or all-dominant way. Rather, reflexivity exists at this level.

An answer to the challenge: the roots and the new flowers

From Schumpeter I to Schumpeter III

A more localised approach, taking its starting point in singularities and alternative strategies, is important in order to understand how firms respond to the general condition of innovation and reflexivity. It can also inform us as to how reflexivity can be mastered by the single firm and by employees, both inside but also, and in particular, outside the main stream of innovation and economic development.

A way out of the contextual limitations that follow from a localised app-roach is to focus on the strategies that are constructed at the micro-level. This moves the analytical perspective from the general to the specific without losing sight of fundamental theoretical categories. The concept of strategy involves a subjective as well as an objective dimension. Thus, strategy has to be manipulated by individuals on the one hand and be socially recognised on the other hand. Furthermore, it establishes an analytical category for one of the units that make it possible for actors to change their situation.

The challenge is to find a Schumpeter III explanation (cf. also Chapter 3 by Gallouj), that describes innovation as strategic reflexivity, where reflexivity is the general condition to which a variety of alternative strategies and role performances exist as responses.

The understanding of innovation that this book argues for goes back to the roots of innovation theory, namely Schumpeter's theory of economic development (1934). However, it elaborates on Schumpeter's original explanation of economic growth. Innovation is still the central category, but the creative process of innovation is less guided by the inner-directed fighting power of individuals that Schumpeter emphasised in his entrepreneur theory. Nor is innovation seen as guided by rationally planned R&D departments or other institutions of science and technology. Innovation is a social process where actors manipulate with and perform strategies and roles. Innovation takes its starting point in a strategic interpretation of the reflexive market or demand.

Innovation is seen as a complex, anarchistic and unsure social process containing many alternative possibilities. Innovation is a game where the players become occupied with defining strategies and roles for themselves and beat other players. They are creative players, thus they invent not only new clever moves, but also new, or differing, rules of the game. Each of them develops their own strategy (cf. Chapter 4 by Sundbo in this book). Firms do not behave unanimously in smooth networks or innovation systems or in relation to one and the same selection environment. They construct alternative strategies within a general condition of reflexivity and make all the alliances they can to win the game. Economic development becomes, in the service economy, even more dependent on actors' steps; it can no longer be framed by dominant strategies (cf. also Chapters 9, 13 and 14 by Gustafsson, Edvardsson and Sandén, Borras and Jæger). Also internal actors are crucial for the innovation process as Meeus and Oerlemans and Matsson demonstrate in Chapters 11 and 12.

Another difference from Schumpeter's original view is that he considered innovation as a periodical activity which, at certain moments, disturbed economic equilibrium (Schumpeter 1939). For the moment, we must consider innovation as a more permanent activity in firms. Furthermore, innovation does not necessarily disturb an equilibrium but may the other way round also be seen as an effort to re-establish an equilibrium (cf. Kirzner 1973). Disturbances and asymmetric chocks are experienced at various levels in the economy, which give rise to various initiatives to adjust and innovate in order to restore a balance between the firm and the environment.

Innovation is taking place within an interactive pattern. Actors and firms have become more interdependent and innovations more connected to each other. What happens in one firm has consequences for another. No firm exists in an independent structure. The entrepreneurial role also changes. Entrepreneurial activity becomes characterised by the Kirzner-entrepreneur. The Kirzner-entrepreneur is the co-ordinator that tries to restore a balance or equilibrium and respond to changes in the environment.

Reflexivity

Contemporary sociology has introduced the idea of a reflexive society (Giddens 1991, Beck 1986, Bourdieu and Wacqant 1992). According to this view, people in modern societies are not inner-directed, for example by an ideology or by fulfilling a psychological or basic physical needs. Furthermore, people's roles in society are not given by tradition or fate. Nor do they just follow the crowd as in the old industrial societies. People follow their individual trajectory, but the world is complex and full of possibilities and dangers. Therefore, people in modern societies inevitably reflect upon their situation. They think about themselves in a radical sense. They also try to calculate what would be clever to do. It is in this sense, people create their own personal strategy (cf. Giddens 1990). Since social relations and infrastructure become more complex, people cannot know everything about what they do. They have to act under uncertainty and rely on experts for help.

Some sociologists focus on the negative sides of that development: the problems of dealing with risks (particularly Beck) that can no longer be calculated or compensated for, how people become confused and how they become dependent on experts. To explain innovation, we must also focus on the possibilities that modernity presents. The risk and uncertainty must be treated from a positive perspective: innovation is a way to develop solutions to problems of firms and individuals and to reduce the risks. Strategic reflexivity is also more generally a situation where firms and people, more independently than just a few decades ago, try to find solutions by themselves. They are not fixed in roles and structures in the same way. This is not the paralysing negative perspective that for example Beck could be said to launch, but an action oriented one: strategic reflexivity is a condition where people can do something about problems through innovation.

Of course, strategic reflexivity also has to produce profits for the innovating firm. One might discuss whether the world's problems may be solved within a non-market economic system (such as a socialistic one as Schumpeter (1943) discussed) or in non-market social systems (such as Habermas's (1987) lifeworld). Whatever the opinion, such questions are outside the scope of this book. However, we do not want to provide a rational, functionalistic view of innovation. We will not argue that innovations are always accepted by the market because they solve specific problems for individuals or communities. A large proportion of the ICT revolution has produced redundant innovations, for example. Innovation does not solve all problems, it is not a linear process which leads directly to the goal. Often it creates more problems than it solves. Innovation contributes to solving some problems as experienced and interpreted by people. We cannot predict which innovations will solve what problems.

Barriers to Schumpeter III

How can the firm survive in such a world? It must be able to make interpretations and make choices, recruit personnel who are engaged in critical dialogue, establish reflexive roles and change the relation to the environment to become more complementary and flexible. Indeed this is not easy, since it represents a radical shift from earlier job relations and industrial relations.

Strategies have to be clear and continuously maintained at the macro-management level in order to create roles for employees and customers. Roles have to be institutionalised at the micro-management level in order to carry out and maintain the strategies in practice. While employees become more independent in order to occupy these roles, they probably also become less loyal to the firm. Therefore, they have to be offered new career opportunities or new forms of barter. In addition to this, the employees' motivation has to be changed. Employees cannot have a conventional instrumental relation to their job (as a place to earn money). Their job relation should become one of personal development as well. But who wants to be 'personally developed' by/at their job?

We must remember that for the firm the important issue is not innovation per se, but to continue to maintain an economic position and a role in society. This is, essentially, what drives the firm, and employees. Whether or not the employees appear to be happy with the job, the job is still a social position that affects the possibilities and prestige of the single individual.

Innovation is therefore often bottom-up and consists of continuous processes of adaptations and changes in service-roles. Furthermore, in the service economy, changes of workers' performance strongly affect the single employee as he or she has a relation to the customer and may be personally blamed for a malperformance. In the service economy, customer expectations may often block innovations at this level, since a conservative attitude to customers' expectations may seem safer during service production. These phenomena are also known in the industrial sector, as demonstrated by Trott in Chapter 7, but here, the single employee does not carry responsibility in the same way.

Some firms are able to generate more radical innovations. They often come in clusters. The most important example is the ICT industry. However, the ICT firms have only achieved success because they are highly valued on the stock market. Lately, the investment bubble in the

ICT sector has been much debated in the media. Thus, in this case radical innovation turns out to happen under conditions of reflexivity as well. The firms, whether they want it or not, have to position themselves in relation to expectations at the stock market. Innovation is not achieved by a single company, getting an idea for a new product, carrying on with that idea without confronting it with the external expectations and customer reactions. This again makes changes much more complex than imaged by Schumpeter in his original explanation of innovation, and strategic steps much more important. Strategic responses to expectations become critical to innovation: what we term strategic reflexivity.

Innovation and expectations about the future

Innovation has to be initiated based on certain expectations about the future, since innovations are realised only in the future, and nobody can know exactly what the future market will look like.

Market seizure

In terms of economics, the firm's innovation behaviour must be understood as market seizure. The firm attempts to position itself on the market. This can be done in two related ways – first and second order seizure. *First order seizure* is when the firm makes itself conspicuous by introducing new products and services or lowering the prices on existing products through productivity increase. *Second order seizure* is when the firm seeks to set the agenda for what the market is interested in, for example new technology or new services such as electronic commerce.

Second order seizure is closely related to expectations and strategic reflexivity. It is, for example, difficult for potential investors to evaluate what the firm is capable of doing. Evaluations will, to a large extent, have to rely on an impressionistic assessment of the firm's competencies and opportunities. The input into the innovation process is, apart from capital, constituted out of soft factors such as creativity, knowledge and social intelligence that are difficult to measure. The importance of intangible factors such as routines (Nelson and Winter 1982), human resources (Teece and Pisano 1994), and knowledge (Nonaka and Takeuchi 1995, Faulkner and Senker 1995) have been widely acknowledged in the literature.

Previous success of the firm is also difficult to measure and apply when the firm seeks to position itself. Previous success could be measured by increases of turnover and profit. Nevertheless, it may be difficult to estimate how many changes in turnover should be credited to what innovation strategies. Furthermore, success in one area of innovation cannot easily be transferred to another area since the next situation may require different skills.

The firm is strongly dependent on expectations from the environment and its ability to deal with these expectations. Since innovation is risky and the outcome unsure, the firm will often try out several innovation strategies at the same time. To use a metaphor, the firm is fishing – it sets out more hooks than it expects to catch fishes. It cannot know what hooks will catch the fishes.

Information about innovation cycles or market analyses may tell the manager when it is time to accelerate the innovation activities, but the firm may successfully do that at another time as well. It may also experience failure in both cases as well. Since there are many possibilities, the firm has to develop strategic reflexivity that takes into account the reactions and expectations of the environment and enable the firm as well as the environment to act with incomplete information.

The manager has several possibilities for choosing the best strategy. She can listen to the R&D department, to market research, to experts and to employees. Attempts have been made to set external criteria for the choice of strategy by emphasising a few market situations and a few strategies related to these situations (cf. for example Porter 1980, 1990, Ansoff 1982). But today, such an approach is more difficult due to globalisation. For one thing, the home market is not as important as it was just a few decades ago. Nor are the vertical links to the home industry so important any more.

Management manages the expectations by searching for knowledge about others' expectations, externally and internally. Management actively creates expectations in order to draw the environment's attention to specific problems that the firm intends to solve. The firm also follows the strategic behaviour of the competitors in order not to invest more than necessary in innovation to win the competition.

Expectations generally attach importance to the potential reactions of the environment, particularly those of experts, investors and customers. Political reactions are also important to the firm as reflections of broader social expectations and choices.

Expectations and organisation of innovative activities

Since strategic reflexivity is guided by unsure expectations about the future, it leads to a varied and flexible organisation of innovation activities. On the following pages we will examine some of the central factors relevant to the formulation of strategic reflexivity and briefly discuss how the single chapters of the book will deal with these aspects.

Market expectations

The market is where the game is fought. This is sometimes forgotten in innovation theory although it has been the object of different analyses and theoretical discussions (e.g. Kamien and Schwartz 1982, von Hippel 1988, Nyström 1990). Markets constitute a structure of expectations with which the firm is confronted.

The structure of market expectations has changed, however. Enterprises have started to focus much more on the individual customer's expectations. New production methods have been launched, such as flexible specialisation (Volberda 1998), modularisation and mass customisation (Pine 1993, Sundbo 1994), that makes it possible to produce customised products. Furthermore, customers' intellectual and economic resources have increased, making them more demanding and focused on quality aspects.

The need to manipulate market expectations and customer roles is greater than ever because people have more resources than ever. The results are also more unpredictable because customers are more critical and independent. Luxury needs or new, advanced needs are appearing (such as the need for information or personal, mental and physical well-being).

Firms must find a method to 'read' market expectations and use that as a basis for their innovation activities. Possibilities and problems are treated in some of the chapters. Nyström and Liljedahl analyse, in Chapter 5, how firms combine the market and technology perspective in their innovation behaviour. Gallina examines, in Chapter 6, how market conditions set the framework for small and medium sised enterprises' internationalisation. Trott's analysis, in Chapter 7, demonstrates how too much traditionalism in firms' interpretation of market possibilities can be an obstacle to innovations. Gustafsson, Edvardsson and Sandén illustrate how the customers become the core of firms' innovation thinking (Chapter 9).

Strategic manipulation of expectations

The firms' approach to market expectations is strategic, not objective. Firms make as many analyses as they can. They then have to conclude how they want to use these analyses. They develop a strategy, which provides a guideline for its management of expectations, and what ideas to qualify and what to disqualify. Innovation strategy is a core part of the technique, through which the firm expresses expectations to future innovations. Innovation is thus strategic reflexivity rather than reflexivity as such and management of innovation must be considered a strategic discipline (cf. Tidd, Bessant and Pavitt 1997, Trott 1998).

Strategic reflexivity may be internally as well as externally oriented. Strategy can be a way to mobilise managers and employees of the firm for specific purposes and ensure a role for them (e.g. Pettigrew and Whipp 1991).

In this book, the internal strategic aspect of innovation is especially examined in Chapter 4 by Sundbo, Chapter 5 by Nyström and Liljedahl, Chapter 8 by Bessant, and Chapter 10 by Fuglsang.

The political manipulation of expectations is also discussed. The state can no longer create economic growth by launching large development programmes. Nevertheless, it can formulate new policies, which co-ordinate market expectations and broader social and political expectations to innovation. This means that the policy becomes more indirect, an attempt to formulate strategic reflexivity, stimulate a co-ordination of expectations, and involve relevant actors in negotiations. This is demonstrated in Chapter 14 by Jæger.

Strategic reflexivity as controlled chaos

The arguments above lead to a discussion of how the innovation activities are organised. In the words of James Quinn (1985) innovation is 'controlled chaos'. The process is unpredictable (cf. Kline 1985). On the other hand it is possible to manage chaos so as to generate an outcome to the firm – even if it is not possible to tell which outcome will be the most profitable (Stacey 1993, Quinn 1985). Strategic reflexivity can be seen as an instrument for managing chaos.

Managers' choices can be deduced from the environment's expectations or from those inherent to internal resources and roles. In this introduction, it has been argued that the latter is insufficient because the environment's expectations is a crucial factor. Furthermore, lack of internal resources can be compensated for; routines can be changed.

If chaos becomes too controlled, it will injure the innovation process (as Trott's Chapter 7 argues) and if it becomes too chaotic, innovation will be injured as well. That is one reason why firms must take the conflict or consensus of internal stakeholders into consideration as Meeus and Oerlemans' Chapter 11 illustrates.

Bessant analyses the issue from another side in Chapter 8. He emphasises routines as a tool that ensures stability in the innovation process, and which makes management of the chaotic innovation process possible.

Postmodern management of expectations

The firm's innovation behaviour may, from the perspective outlined above, be studied in light of some of the recent postmodern discussions within sociology and organisation theory (e.g. Kumar 1995, Giddens 1990, 1991, Hatch 1998). The term postmodern may be a bad one because it is semantically self-contradictory. However, the viewpoints in the discussions are relevant to explain the situation of contemporary organisations and firm behaviour: a post-fordist, flexible organisation (Volberda 1998) is emerging with individuals of different norms and behavioural patterns. In Chapter 2 by Nyström the postmodern perspective of the innovative organisation is discussed.

Further, individuals are to a greater extent relating their working life to their social life outside the job, which means that the border between work and leisure becomes more fluent. This also has implications for the motivation of employees (as discussed by Kanter 1983, 1989), which in some (but not all) cases becomes higher. Employees also become more aware of the borderlines they want to draw between themselves and their job. More loosely coupled organisations (such as networks, virtual organisations etc.) emerge (cf. Chesbrough and Teece 1996, Sundbo and Gallouj 2000). A variety of organisational types and individual behaviour within organisations occur.

This happens not just because environmental conditions change, as contingency theory says (Lawrence and Lorsch 1967), but also because individuals bring different norms and perspectives from their lives into the organisation. Organisational culture (e.g. Schein 1984, Hatch 1998) becomes a core factor, also for innovation.

Management becomes a central factor in innovation. Active management is required to create an innovative recovery movement. The organisation will not do it by itself, nor will any knowledge and technology based innovation system. Sometimes the market situation is good, but often it is unpredictable. Because firms and political authorities take a strategic approach, management becomes central in that situation. The top managers have to make decisions, but based on the market possibilities. They are dependent on a creative firm organisation to develop strategic reflexivity.

The management becomes reflexive, which means that managers seek advice and knowledge and create their personal strategy for providing the foundation for a strategy of innovation. Many actors become involved in this process. Despite an increasing individualism, the system is held together as much as ever.

Knowledge is, however, not sufficient. The firm must act in new ways, it becomes *strategically* reflexive, and creativity becomes important. Creativity is individual as well as organisational. Both require management. This is, among other things, argued in Chapter 5 by Nyström and Liljedahl concerning open management strategies.

Innovation as an interaction process

Innovation in the firm is a social and organised process. Since innovation is so complicated and the managers are so uncertain about what to do, innovation becomes a broad, social interaction process in which many parties participate. The employees and managers participate in different ways (cf. Burgelman and Sayles 1986, Kanter 1983). The organisation can be more or less tight or loosely-coupled and it can be carried out by special departments such as an R&D department. Many employees and managers who are not specialists can be involved. Bessant discusses the general innovation system and its challenges in Chapter 8.

The innovation process has different phases. In a first 'idea' phase, widespread entrepreneurship from all employees plays a crucial role (cf. Pinchot 1985, Drucker 1985). This phase may also take place in the R&D department (Freeman and Soete 1997). The later phases are generally more project oriented (Cooper 1988, Heap 1989). However, there are many variations. These organisational principles are discussed in Chapter 4 by

Sundbo. Chapter 10 by Fuglsang demonstrates how employees achieve different roles in the process.

Also, external parties are widely used. The employees and managers interact with external actors and use external knowledge bases. As has been demonstrated, firms participate in informal and formal external networks (Håkansson 1987), also in their innovation activities. This is also discussed in Chapter 3 by Gallouj and Chapter 4 by Oerlemans and Meeus. The firms use customers as partners or objects in their innovation processes (Lundvall 1988). This is demonstrated in Chapter 9 by Gustafsson, Edvardsson and Sandén, who illustrate how this is particularly the case in services.

Innovation also takes place in the society outside firms. Innovation processes in society can be seen as diffusion processes, for example of new technology or new social ideas (cf. Rogers 1995). Thus, they are also interactional processes. Such processes are also important, even when we try to understand firm innovation processes, because many of these social innovations may be related to or transformed into commercial activities. That is what is happening to the use of the Internet which started as a social process and is now being transformed into business processes. Jæger's Chapter 14 shows how such social diffusion processes go and which difficulties the processes may face.

Part I

Theory

The first part of the book is theoretical. From slightly different theoretical perspectives, the authors attempt to give more substance to the concept of strategic reflexivity. Many theories of innovation take their starting point in either the concept of the entrepreneur or in technological change – as driving forces behind innovation. In the approach of strategic reflexivity both the role of the entrepreneur and the importance of technological changes are recognised. However, strategic reflexivity is not a determinist concept, but a dualistic one. It means, as it will become clear, both being strategic as well as reflexive. Strategic reflexivity seeks to emphasise the crucial function of creativity, interaction and interpretative strategic processes – and order as well as chaos.

Harry Nyström develops a theory of creative management, using the literature on 'postmodernism' to qualify his notion of creativity. Modernism is, according to Nyström, characterised by convergent processes of change and postmodernism by divergent processes. Nyström defines the creative process as a balancing act between divergent and convergent processes. A divergent process is a characteristic of the early stages of creativity and convergent processes of the later stages of the creative process. The most crucial issue in a creative process is therefore to switch between the two opposing positions. Applying this framework to management theory, Nyström then makes a distinction between economic management (an approach where modernism is dominant) and creative management (combining modernism and postmodernism). In creative management, the requirements of both the early and later stages of the creative process are balanced. Innovation management is seen as the objective of creative management and as major mechanisms for constructing strategic reflexivity.

Faïz Gallouj develops a new model of innovation, the consultant-assisted or interactional innovation model. The purpose is to rehabilitate the role of services in the theory of economic dynamics. Gallouj argues that the interactional innovation model leads to a more flexible and open definition of innovation. Gallouj's model is based on a Schumpeterian tradition but includes services by stressing how services (consultants) contribute to innovation. The model is presented as a Schumpeter Mark III model.

According to the author, Schumpter's own two models are science and technology push models. The Mark III model is inspired by an evolutionary approach, where innovation is a problem-solving activity of a more general character, covering all kinds of problems that exist in a firm – social, legal, fiscal, marketing and strategic issues. The paper presents consultancy as a special kind of problem-solving activity that can relate to all the functions of a firm. The author goes on to present various configurations of the model, including a standard model. Gallouj also speculates about whether interactional innovation will lead to convergence of innovation systems or variety; he gives some interesting arguments to support the variety hypothesis. Other 'dangers' are also considered, such as the risk of cognitive lock-in and appropriability problems.

Jon Sundbo develops an approach to innovation where the strategic process is the explanatory category. In Sundbo's framework, strategy implies both an external aspect, the firm's market position, as well as an internal aspect, the organisation of the firm. In this approach, strategies have to continuously adapt to changes in environment and cannot be generic. The organisation of the innovation process is considered to be a dual one. On the one hand, there is, in an innovative firm, a loosely coupled interactive structure, in which ideas and development of innovations take place in an interactive manner, and on the other hand there is a hierarchical managerial structure, which regulates the innovation process. Strategy formulation is linked to the concept of what Sundbo considers to be a 'Kirzner-entrepreneur' (after Kirzner 1973). In this approach, firms make many adjustments to underutilised market opportunities, not as single entrepreneurs, but as a result of social organisational processes. These small adjustments or changes not only restore market imperfections, they also sometimes create new market possibilities that did not exist before. Sundbo's approach is illustrated by two of his case-studies namely a small bank and an engineering consultancy.

2 The postmodern challenge

From economic to creative management

Harry Nyström

Introduction

In this chapter, strategic reflexivity and strategic management are discussed in relation to modernism and postmodernism. Strategic management, as a planning approach to dealing with reality, presupposes predictability and rational decision-making, the modernistic worldview of economic theory and economic approaches. In this world, strategic reflexivity is a calculated and objective response to environmental change. Innovation in this framework may be seen as the logical result of preexisting conditions, rather than as creative surprise leading to radically new and different experiences and activities.

We may term strategic management, when used as an innovative method, 'creative management'. This may be viewed as a wider and more dynamic approach to understanding and changing reality than that of the economic approach. It not only analyses existing conditions but also constructs new ones. Strategic reflexivity then involves both divergent activities, creating a potential for knowledge, and convergent activities, establishing knowledge. Postmodern thinking is mainly concerned with divergence, which characterises the early stages of the creative process, while modern thinking emphasises convergence, which dominates during its later stages. Consequently, to understand strategic reflexivity and strategic management in an innovative context, we need to combine elements from both postmodern and modernistic thinking.

Modernist and postmodernist perspectives

In the literature, 'postmodernism' has two meanings, one historical and the other epistemological. First, it may be viewed as a stage of development where it follows modernism as a distinct era, involving a new view of the world and a new cultural logic. Instead of emphasising scientific rigor, formal logic, and rationality as a clear basis for understanding and managing society – as is the case in the modern bureaucratic organisation – postmodernism may be seen as characterised by pluralism, fragmentation,

ambiguity and indeterminacy, defying attempts to generalise and extrapo-
late from past experience. On this view, postmodernism marks the end of
the modernisation process and of modern society.

Second, it may be viewed as a challenge to the traditional view of knowledge that assumes that there is a clear way to view and understand the
world. In this second meaning, postmodernism is a timeless phenomenon,
a cognitive style, rather then a time-dependent emerging social trend. In
both cases, however, modernism and postmodernism are viewed as separate and distinct phenomena, with the latter following the former or the
two coexisting, but not creatively interacting over time.

Paradoxically enough, postmodernism – in itself a challenge to a deterministic worldview – is therefore seen as part of a deterministic process
of change, both in its historical and in its epistemological sense. The argument would seem to be that both society and our way of viewing the
world are changing in a postmodern direction, and that this is an inevitable
consequence of development itself.

In contrast to this view of modernism and postmodernism, I would like
to offer a somewhat different interpretation, which views them as part of
the creative process, both on the individual and on the societal level.
Instead of regarding these terms as deterministic, time-dependent development factors, I will view them as global terms for what characterises
the early, versus the later, stages of the creative process. This means that
they are determinants, rather than consequences of change, describing
mechanisms and processes, rather than outcomes.

It also implies a management perspective (Boje *et al.*, 1996, Cova, 1996),
which sees modernism and postmodernism in terms of the ways we can
manage society and organisations, and not only observe what happens in a
historical perspective. Instead, of viewing society or knowledge as more or
less modern or postmodern, this approach implies that it is the management
of society and knowledge which is more, or less, modern or postmodern.

In the historical sense, modernism implies the rationalisation and differentiation of social and economic activities that accompany capitalism and
industrialisation during the transformation from traditional to modern
society (Hollinger, 1994). This is a periodising view of development where,
as in Marx's and Weber's writings, modernisation is seen as the socio-
economic outcome of a process of change. The main thrust of the argument
is that modernism is different from what precedes it, which, depending
on our point of view, may be seen as beneficial or detrimental to either
society or its constituents.

As a philosophical concept, the essential issue concerning modernism is
the nature of knowledge and the methodological requirements for determining truth and meaning, developed by the philosophers of the
enlightenment. Logical thinking leads to objective truth in this worldview,
and science is the way to achieve a just and stable society. An objective
account of the world must be guided by objective rules, that is by method.

Predictability is needed as a basis for rational decision-making and both leads to, and is made possible by, stable conditions. It is primarily this meaning of the term that is of interest in the present discussion and it is the assumptions inherent in this view that are most strongly questioned by postmodern critics of modernism.

The term postmodernism literally means after modernism, and may be seen as implying that in some way postmodern conditions have supplanted modern ones. In the historical sense, depending on our frame of reference and what variables we focus on, postmodernism may be viewed as either a radical break with or continuation of modernism (Cummings, 1996). This, however, basically depends on whether we focus on differences or similarities and since one of the main ideas in postmodern approaches is to stress differences (Staten, 1984, p. 23), the result is that postmodern theory, by its own assumptions, emphasises its break with modernism.

To understand postmodernism we need to look not at reality as such, which modernism claims to do, but to the way in which we construct reality, that is to language as a basis for thought and understanding. Language games, discourses and narratives become the basis for postmodern know-ledge and method (Lyotard, 1986) and meaning is not given but socially constructed. This leads to a diversity of ideas and plurality of postmodern perspectives and positions (Best and Kellner, 1991). Boundaries between academic disciplines and between theory and practice dissolve, and fragments from different sources are combined into idiosyncratic maps of society. In the extreme versions of postmodernism, the distinction between what is real and what is unreal tends to disappear in the most extreme versions of postmodernism, where images replace substances in what is called hyperreality (Baudrillard, 1983).

The creative process and postmodern theory

In this section, we will compare the assumptions of postmodernist theory with some of the basic ideas presented in the creativity literature (Isaksen, 1987, Sternberg, 1988). The similarities between these two ways of looking at the world are striking, but surprisingly enough, this has been seldom noted by writers in the different traditions. Writers on creativity, with the exception of Rickards (1999), seem to have paid little attention to post-modernism as a term or philosophical movement and postmodern writers seldom use the word creativity, or make any references to the vast liter-ature directly dealing with this phenomenon. This is all the stranger since writers in both these areas often stress the need for interdisciplinary approaches in scientific work, for instance Lyotard in his influential essay *The Postmodern Condition* (1986, p.52.) Or Koestler in his pioneering book *The Act of Creation* (1964, p.164).

How then may we understand creativity? Bruner's term 'going beyond the information given' (1974) is an interesting idea, particularly today in our computer society and information age, as is his term 'effective surprise (1974, 18ff.). Another interesting phrase is 'making the strange familiar and the familiar strange' (Lincoln, 1962). Koestler (1964, p. 35) speaks of creativity as the biosociation of different matrices which certainly captures an important aspect of the linking together of previously unrelated elements of knowledge. From a psychoanalytical perspective Arieti defines creativity as the magic synthesis of conscious and unconscious thought processes (1976, p. 13).

Common to all these notions seems to be the idea of creativity as something unexpected and surprising, emerging from outside the realm of rational thought processes, which is seen as the playground for modernistic approaches to theory generation. It is a question of calling forth and realising the unknown. In Lyotard's words, (ibid., p. 81) to 'invent allusions to the conceivable which cannot be presented', an activity which he views as a main characteristic of the postmodern world. This would seem to be related to the psychoanalytical notion, used in the creativity literature, of a crucial incubation period in the creative process between preparation and illumination (Wallas, 1926). This is when unfocused and dreamlike processes are transformed into definite formative ideas in a struggle between depth and surface perception (Ehrenzweig, 1967).

It is by looking at the creative process itself that we most easily recognise the similarities and differences between postmodern views of knowledge and creativity theory.The creative process (Nyström, 1979, 1990) may be seen as a balancing act between divergent and convergent thought processes, in order to generate new knowledge and develop original and constructive solutions to perceived problems. Usually, the process is seen as consisting of several stages. The early stages are largely dependent on divergent processes for success, while the later stages are highly dependent on convergent ones.

The most crucial issue is that of switching constructively between the opposite processes, and this is complicated by the fact that some individuals are more skilled at or inclined towards divergent thinking and others towards convergent thinking. More open cognitive mechanisms such as intuition and visual thinking are more important during the early stages, while more closed cognitive mechanisms, such as formal analysis, are better suited to the requirements of the later processes. Tolerance of ambiguity and intellectual flexibility, as well as openness to experience and a willingness to experiment, are other abilities that are advantageous during the early stages.

Differentiation (Lasch, 1990) can be seen as a major characteristic both of the later stages of the creative process and of modernity and dedifferentiation of the early stages and of postmodernism. In the psychological sense (Witkin *et al.*, 1962) differentiation implies making detailed comparisons, distinctions and judgments between different entities. This requires formalisation and should tend to increase clarity, a major requirement for analysis and systematic discourse, the two being important during the later stages of the creative process. Dedifferentiation, on the other hand, by relaxing perceptual boundaries (Arieti, 1976, p.44) leads to an increase, rather than a decrease. This, as we have noted above, is important in order to reduce psychological closure during the earlier stages of the creative process.

Postmodernism, by stressing dedifferentiation, would appear to be open to the same type of criticism that may be directed towards early creativity theory, namely that too much focus is placed on the early, more open, unformalised stages, and too little on the later more closed, systematic stages. This misses the main point that creativity is the result of interaction between more unstructured intuitive judgment, and more structured analytical reasoning. Using Wittgenstein's terminology, we may say that new insight and creativity depends on the interplay between different language games or discourses, something which only recently has been recognised in the creativity literature, and is still not sufficiently stressed in postmodern approaches to knowledge generation.

Deconstruction is one of the most used and least precise terms in the literature on postmodernism. Making allowance for this, the above mentioned definition of creativity – making the strange familiar and the familiar strange – still seems to capture some of its intended meaning. More directly, deconstruction may be defined as a regulated overflowing of established boundaries (Staten, 1984, p. 24.), and a rethinking of accepted meaning. However, this rethinking may be either used primarily in a negative critical capacity, dismantling reality, or seen as a possibility for constructive change. In the postmodern literature, it would seem to be the former meaning of the term that is emphasised, while in the creativity literature the latter meaning of the term takes precedence. At the same time, critical capacity, by going beyond the information given and uncovering hidden content, may be seen as a necessary but not sufficient condition for creativity. Postmodern theory, however, seems to be more concerned with deconstruction as such, than with using it as a basis for creative reconstruction. Other terms used to describe postmodern approaches in the literature (Cooper and Burrell, 1988) are, for instance, problemising, edifying discourse, images, disorder, indeterminacy and heterogeneity. All of these terms could equally be used to describe the early stages of the creative process, but the literature tends to view them as isolated descriptions of existing or preexisting conditions, rather then as part of a constructive future oriented change process.

Problemising is questioning rather than verifying, the latter activity being given priority in modernistic approaches to knowledge generation. Edifying discourse (Rorty, 1980, p. 12) helps us to break free from existing notions and ideas, compared to using systematic discourse based on logic and the prevailing reason which may be seen as a limiting and stabilising factor in modernism. Disorder, indeterminacy and heterogeneity may be viewed as characterising the postmodern condition, leading to ambiguity and unpredictability.

Images, in postmodern theory, tend to be viewed as more real than the reality they are supposed to mirror (Baudrillard, 1983). Instead of reflecting reality, they constitute it. Essentially, however postmodern writers see images as being the result of erratic uncontrollable development, rather than as determinants of constructive change which may be managed indirectly by creating favorable conditions. In a more creative framework images, as a result of what we may call their constructive vagueness (Nyström, 1990) may make possible both flexibility and direction. Thus, they can facilitate constructive change during the early stages of the creative process, when intellectual flexibility is needed, and ambiguity and unpredictability make it impossible to use logical analysis to determine what is best to do.

Image is also related to intuition (Bastick, 1982), another central term both in the creativity literature and in postmodern thinking. Croce's view of intuition as the 'undifferentiated unity of the perception of the real and of the simple image of the possible' (Aieti, 1976, p. 408), presented in 1909, would seem to be a precursor or forerunner to many of the central ideas in postmodernism. At the same time, intuition can be seen as a creative mirror, both reflecting and transforming reality. By not faithfully reproducing reality images may be viewed as innovations. This again points to a certain parallelism in ideas between creativity theory, where images may be seen as predeterminants of creativity, and postmodern thinking, where they are the basic substance constituting reality and therefore also the medium for possible change.

Economic management

Economic management is based on the worldview embraced by economic theory. It is a relatively closed and static view, emphasising stability, continuity, homogeneity and fixed ways of organising resources to achieve short run efficiency in producing and marketing products. It assumes rationality in both expectations and decision making, objective knowledge, reducibility and single value optimisation. This is much the same as the worldview of modernism, characterised by clarity, order, predictability, rationality and systematic discourse.

This is a model of human behavior and society that owes much of its intellectual heritage to Hobbes (Hollinger, 1994). Using physics as a basis for his thinking Hobbes viewed human behavior as predictable and controllable, governed by rational choice. Based on free exchange between parties, cost-efficiencies result as 'reasoning becomes reckoning' (ibid., p. 22). Following Hobbes, Adam Smith went even further, turning rational self-interest into a moral virtue by postulating an invisible hand. He used pure competition as a theoretical device to resolve the conflict between individual self-interest and societal well being. Even if we accept the underlying modernist assumptions of this thinking (human behavior is rational and based on complete knowledge of an objective reality) the restrictive assumptions with regard to how society and the economy function make its difficult to find applicable real world situations.

Differentiation is perhaps the best term to summarise the economic approach to management, and this, as we have noted above, is also one of the best terms to describe modernism. Differentiation requires clarity and leads to order in describing and managing the world, and therefore may be seen as the main analytical tool in modernism and economic management. It is the basis, for instance, for strategy formation found in the economic management literature where strategy formulation is seen as an analytic process (Ansoff, 1965, Porter, 1980).

At the same time, complexity and diversity make it difficult to differentiate the world according to the assumptions of economic theory and we therefore need to simplify the model to make it work. Assumptions such as identity between competing products, continuous aggregate demand and cost functions make it possible to arrive at determinate solutions to the optimisation of individual and societal welfare. However, it is difficult to find, or even imagine, real world situations that fit these requirements. Complexity and heterogeneity, rather than simplicity and homogeneity, are used by most insightful observers to describe the world, and these are conditions that are not easy to deal with analytically in modernistic socio-economic models.

The assumptions of economic management are also similar to the requirements of the later stages of the creative process, where formalisation is used to achieve analytical rigour in testing and implementing the insightful solutions derived from the earlier stages of the creative process. The basic assumptions of a modern world view and economic theory therefore should not be viewed as useless for understanding and managing a world, which seems to correspond more closely to the postmodern than the modern view of reality, as some postmodern writers would seem to imply. Instead, they should be seen as applicable to different stages of the creative process in the construction and implementation of the world we live in and to different types and levels of activity.

Entrepreneurship, for instance, which places action before analysis in strategic reflexivity, is best understood against the background of creativity theory (Nyström, 1993) and postmodern theory, while day-to-day decision-making in established firms is closer to the modern model of how companies and the economy functions. Regarding society and the economy as essentially modern or postmodern, therefore seems not to be a fruitful way to achieve a better understanding. Instead, elements of both these approaches should be combined to achieve a more adequate view of the world today.

Creative management

Creative management (Nyström, 1989, Henry, 1991) is a more open, dynamic and flexible approach to management than economic management and includes economic management as part of its framework. In this approach, the requirements of both the early and later stages of the creative process are balanced. The early stages may be seen as postmodern and the later stages as modern, but the focus is on the total process of change, rather than its components.

It is a dualistic management approach, combining opposing tendencies to achieve a balanced outcome. Creative management considers differentiation and dedifferentiation, continuous and discontinuous change, objective and subjective knowledge, intuition and analysis, predictability and experimentation, reducibility and holism. It may be seen as an integrated economic and psychological approach to product and company development (Nyström, 1979, 1989).The early more postmodern development stages emphasise intuition and entrepreneurship, both enacting and developing visions and managing images. The later, more modernistic stages emphasise systematic analysis and planning in order to reduce cognitive uncertainty and establish order.

Consequently, image and innovation management may be seen as the main overall objectives of creative management and as major mechanisms for constructing strategic reflexivity. Products are the outcome of the managerial process, as the word itself implies, and producing products therefore mainly reflects the requirements of the late stages of the creative process, viewed as the way from idea to market offer. Product innovation and marketing, however, are much more concerned with the early stages of the creative process.

Different buyers and consumers usually view, and use, products in different ways. As a result, images become better predictors of product demand than products, described in terms of their physical attributes. This, of course, is particularly the case when products are difficult to evaluate according to objective, a priori criteria. Non-standardised services may be seen as one type of product where this is the case, but also products such as jewellery or oil paintings, which are bought more for their subjective linkage to personal values, than their objective functional use, as in the case of petrol or stoves.

Image marketing (Nyström, 1989, Cova, 1996) is more concerned with the psychological reality of buyers (which postmodern theory focuses on), while product marketing is more concerned with their objective needs stressed in traditional economic theory. If the emphasis is on image marketing, focusing on subjective psychological values, postmodern theory is more useful as a guide to understanding the relevant reality. If the main concern is product marketing, stressing more objective physical and technical attributes, traditional modern management ideas are more applicable.

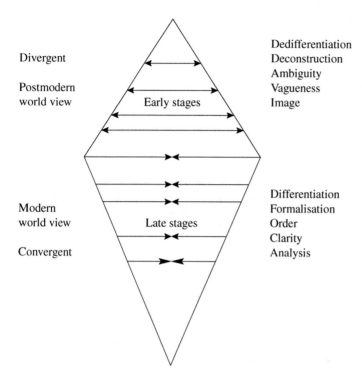

Figure 2.1 The creative process

However, to understand the whole process from developing new products to producing and marketing them we need to combine postmodern and modern thinking.

Combining modern and postmodern management

Ideally, we need to combine modern and postmodern management styles to achieve constructive creative change. Traditional economic theory and economic approaches to management, I will argue, are based on a modern worldview and mechanisms for realising this ideal. This reflects the needs during the later stages of the creative process, when applying existing knowledge is the main concern. In a rapidly changing world, with a greater need for developing new knowledge, we need, to pay more attention to the earlier stages of the creative process, which may be seen as corresponding more closely to the postmodern world view.

The crucial requirement, however, is to integrate and balance the early and later stages of the creative process. Instead of viewing postmodernism as the end of the modernisation process, I suggest that it should be viewed as the beginning of the creative process, that is, the creative transformation of society (see Figure 2.1).

In a creative framework postmodernism, as a set of conditions, can therefore be either pre- or postmodern, depending on the timing. That is, whether it is the beginning of a new creative cycle or the end of an old one. In a dynamic development perspective these cycles repeat themselves over time, as creative destruction (Schumpeter, 1934) alternates with creative construction.

This, then is the postmodern challenge to management on all levels of society. We need to move from a traditional, modern economic management approach to a creative management approach by integrating modern and postmodern elements, rather than focusing on one to the exclusion of the other. These elements have always been intuitively balanced in practice by insightful decision-makers, but are separated in theory. A creative management approach should therefore also help to bridge the gap between academics and practitioners in describing and understanding their relevant realities.

Summary

To understand strategic reflexivity and strategic management in an innovative context, it is helpful to contrast and bring together ideas from both economic theory and creativity theory, seen in relation to modern and postmodern interpretations of reality. Economic and creative management may, then, be regarded not as two different ways of understanding and managing the world, but as part of an integrated framework. This is done against the background of the creative process, where the early, more open and flexible, stages basically reflect the requirements of postmodern theory. The later, more closed and analytical stages, are similar to the modern worldview of economic theory and economic approaches to management. In a rapidly changing world, where we need to both develop and apply new knowledge, we should adopt a balanced modern and postmodern approach to theory generation and implementation, to understand and evaluate both individual, organisational and societal action.

3 Interactional innovation

A neo-Schumpeterian model

Faïz Gallouj

Introduction

Innovation in services, long neglected by economic theory, is currently becoming an established area of inquiry. This could hardly be otherwise in economies in which services are increasingly important in terms of their contribution both to employment and to value added.

The growing interest in this issue is putting an end to a paradox: that of contemporary societies whose two main characteristics are said to be innovation and services but in which the service sector is, allegedly, impervious to innovation.

The explanation of this paradox is rooted in the origins of political economy.[1] Its analytical tools were conceived *in* and *for* industrial and agricultural economies. Adapting these tools to the specificities of services, and particularly to their often intangible and relational nature, produces two significant results. First, innovation in services is shown not only to exist, possibly in a variety of forms, but also to play a not insignificant role; second, various possible means of enhancing the ensuing industrial innovation are illuminated.

The purpose of this chapter is to advance what might be called the 'rehabilitation' of the role of services in economic dynamics. Indeed, services (or at least some of them, particularly the most knowledge-intensive) are not simply producers of innovation. Increasingly they are becoming, in different ways, the medium for innovation in other spheres of economic activity.

In other words, this chapter is concerned not so much with innovation *in* services,[2] as innovation stimulated *by* services – although, in reality, as we shall see, these two activities may be inseparable. Indeed, this overlap raises considerable problems concerning the appropriation of the gains from innovation and makes it sometimes difficult to distinguish between the routine activities of service providers from their contribution to innovation.

Recent literature offers a range of interesting analyses of the role played by services in their clients' innovations. There are, for example, a number of studies of the externalisation of R&D activities and of the role of 'public intermediary agencies' in the dissemination of scientific and technical information. Yet studies devoted exclusively to consultancy in information and communications technologies seldom go beyond these issues (Djellal, 1995). Bessant and Rush's (1995), and Hales's (1997) analyses, on the other hand, are concerned with the role of consultants in 'technology transfer', while at the same time extending the meaning of terms a little: here, 'transfer' is not limited to its linear dimension, nor 'technology' to its material dimension. In turn, Miles *et al.* (1994) examine the various roles played by knowledge-intensive business services and describe these activities as users, diffusers and producers of innovation. On the macroeconomic level, Antonelli (1996) uses data from input/output tables and a methodology based upon percolation processes in physics to analyse the role of information and communication technologies in improving agents' 'connectivity' (i.e. the number of connections established between agents in a network) and 'receptivity' (i.e. their capacity to absorb information) through increased use of knowledge-intensive business services. Finally, mention could be made here of the literature that focuses on the reasons for knowledge-based services outsourcing. The argument that outsourcing may produce greater innovative capacity is frequently put forward in support of such a strategy (Quinn, 1992, 1999; Gadrey (ed.), 1992; Albert and Bradley, 1997).

The 'consultant-assisted innovation' or, more generally, the 'interactional or network innovation'[3] model developed here aims at unifying these analyses while at the same time opening up other perspectives – in particular the possibility of extending the range and variety of mechanisms and forms of innovation. Moreover, this model can be shown to be an extension of Schumpeterian and neo-Schumpeterian analyses. Indeed the (neo-Schumpeterian) definition of innovation as an activity leading to the solution of 'selected problems' (Dosi, 1982) fits particularly well with consultancy activities, the purpose of which is precisely to find preventive, curative or creative solutions to problems detected or (re)constructed in collaboration with the client.

The interactional view of innovation is not a new one. It lies at the heart of certain recent and important studies in economics, sociology and management sciences which are concerned in particular with innovation systems, networks, alliances and technological collaboration (Lundvall, 1988, 1992a; Von Hippel, 1988; Hakansson, 1989; Powell, 1990; Callon, 1991; Nelson (ed), 1993; Lamming, 1993; Freeman, 1995; Combs *et al.*, 1996; Edquist (ed.), 1996; Cohendet *et al.* (eds), 1998). The model developed here has the capacity to enhance this general interactional approach to innovation, particularly (though not exclusively) by focusing on the role of services.

Beyond this, however, the proposed model extends, in an interesting and unexpected manner, the two models of innovation developed by Schumpeter in *The Theory of Economic Development* (1934) and *Capitalism, Socialism and Democracy* (1943) and known subsequently as Schumpeter Mark I and II (see Phillips, 1971).

Interactional or network innovation is, as we shall see, a new expression of the Schumpeterian notion of enterprise. In this sense, the model might be baptised Schumpeter Mark III, or the 'interface model of entrepreneurship', to mark its affiliation to the Schumpeterian approach.

This chapter is divided into four sections. The first comprises a summary of Schumpeterian analysis. Next, the Schumpeter Mark III or interactional innovation model, its different components, and its mode of operation are introduced. In the last section some theoretical implications of this model are examined, including the difficulties involved in moving from a microeconomic model to higher levels of analysis (network, system), the risks of convergence between (national) innovation systems to which it may give rise and the risks of technological and cognitive lock-in, as well as the problems of knowledge and innovation appropriation that are inherent in any such model.

Schumpeter and services

There is no part of Schumpeter's work specifically devoted to innovation in services, or in any event to a theoretical analysis of its particular character. Two service activities, however, are repeatedly mentioned in his writings: banking services, considered not so much as innovative activities but as a means of financing entrepreneurial activity; and retailing, which is cited on several occasions as an example of innovation (see, for example, Schumpeter, 1943, pp. 85–6).

In other words, the considerable relevance of Schumpeter's works for the service economy lies more in the potential fruitfulness of his analyses and concepts than in actual studies and results. Particularly useful are:

1   Schumpeter's wide definition of innovation which, with a few adjustments, can easily be applied to services. According to him, innovations are combinations of knowledge resulting in new products, processes, input and output markets, or organisations;
2   Schumpeter's concept of the 'entrepreneur', countless examples of which are constantly being produced by the service economy – which in turn encourages the belief that services may well be the last bastion of 'romantic improvisation' in innovation. In other words, very simple ideas can still lead to the creation of economic empires, as is shown by numerous examples in retailing, catering, hotels and tourism.

These are mere conceptual transpositions. Yet Schumpeter remains useful in other, more subtle ways, one of which provides the starting point of this chapter.

The spirit of enterprise is embodied in two successive (but not mutually exclusive) models developed by Schumpeter: the entrepreneurial model and the monopolistic model.

Let us briefly summarise the main characteristics of each of these models. The central figure in the entrepreneurial model (also called Schumpeter Mark I) is the individual entrepreneur. Endowed with a gift for combining factors in new ways, the entrepreneur acts as a midwife in the process of transforming an (exogenous) invention into an innovation. In the monopolistic model, on the other hand, the locus of entrepreneurial activity is a specialist department (the research and development department in large firms). In both cases, the introduction of the innovation triggers a Darwinian process of 'creative destruction' which undermines existing market structures, to the advantage of innovators and to the detriment of their competitors, who are doomed to fail and disappear. The transition from the entrepreneurial to the monopolistic model involves (at least in part) an endogenisation of innovation, i.e. a process of bureaucratisation that Schumpeter (1943) links with the obsolescence of the entrepreneurial function, and thus, eventually, with the disappearance of capitalism.

The Schumpeter Mark I model, like the Schumpeter Mark II model, can easily be applied to service activities and firms provided that a more flexible interpretation of the notion of innovation is accepted, but one that is not incompatible with the Schumpeterian spirit. Thus, the Schumpeter Mark II model may well characterise parts of the service sector. Although there is rarely an R&D department in the traditional sense, there may exist a set of formal structures devoted to innovation (e.g. flexible project groups whose membership cuts across department boundaries).[4]

Ultimately, though, Schumpeterian analysis can be extended by introducing what we have called the 'interactional innovation model' (or Schumpeter Mark III model, or interface model of entrepreneurship).

Just as the transition from the Schumpeter Mark I to Schumpeter Mark II model characterises, according to certain authors (Freeman *et al.*, 1982), the historical evolution of capitalism, so the appearance of our new model is, to a certain extent, another phase in this evolution. This phase, which Schumpeter could not have anticipated, coincides with the explosion of the tertiary sector. Against this general background, is the advent of a knowledge-based economy, in which 'grey-matter services' or 'complex services' or even 'knowledge-intensive services',[5] constitutes a knowledge infrastructure that complements and competes with the public knowledge infrastructure made up primarily of public education and research services (Bilderbeek and Den Hertog, 1997).

As we shall see, in sum, the advent of this 'interactional innovation model' leads to a more flexible and open definition of innovation. Moreover, interactional innovation, along with other mechanisms (e.g. the implementation of organisational concepts such as the various forms of 'intrapreneurship', i.e. modes of organisation that create spaces in which freedom and creativity, the internal spirit of enterprise, can flourish), has probably counteracted the bureaucratisation of the entrepreneurial function which Schumpeter feared. Indeed, it opens up the firm to the external environment and encourages the renewal of organisational routines.

Schumpeter III: the components of the interactional innovation model

Our model has three components: the innovation and the two actors who produce or implement it, i.e. the 'consultant' and the client. These three terms must be defined before they are brought together and the various possible configurations of the interactional innovation model are examined.

Innovation: definition and objects

As has already been noted, Schumpeter developed a wide-ranging typology of innovation but typically used only a restricted version of it (technological innovation). Schumpeter's two models are, indeed, 'science-push' models. In the Schumpeter Mark III model, a resolutely broad and open definition of innovation, inspired by the evolutionary approach, is adopted. Innovation is defined as a 'problem-solving activity' (Dosi, 1982; Simon, 1989; Egidi, 1997) covering the whole range of a firm's problems (or functions), whether they are considered independently of each other, or in terms of their interactions. Some of these problems are, of course, technological in nature, but there are also social, legal, fiscal, marketing and strategic problems, among others. With the exception of technological problems, the solutions to which are described as product or process innovations, solution of these different problems is generally treated in two different ways by economic theory.

- They are denied the status of innovation, even when the solution or solutions to the problem are original and unprecedented, in order to preserve a degree of 'operationality' in the concept. At most, they are considered as 'change'.
- All (or a large part of) the problems and corresponding solutions are grouped together in a 'catch-all' category labelled organisational innovation.

Schumpeter's typology (product, process, organisation, input and output markets) is useful to escape from this treatment, provided that some semantic adjustments and extensions are accepted. For example, product innovation and process innovation cover both *tangible* products and processes and *intangible* products and processes (e.g. a new type of consultancy or a new field of expertise in consultancy, a new type of hotel or holiday package, a new method, etc.). Furthermore, it is important to enlarge the notion of product innovation, so that it includes a category which we have elsewhere called ad hoc innovation (Gallouj and Gallouj, 1996). This refers to tailor-made and co-produced innovations, as well as novel (and not necessarily directly reproducible) solutions to a given client's problem.

Another way of looking at innovation, which is not incompatible with that outlined above, is one that focuses on the various functions (Fj) within the firm that may be the object of innovative activity. From an evolutionary perspective, a function can be defined as a set of activities (occupations, specialisms or competences) based on a common disciplinary or cognitive field and associated with particular tangible technologies (machines) or intangible technologies (methods). Each of these functions can be associated with its own 'production function'. The different functions in question are well known to researchers in management, whose fields of inquiry and research programmes are, to a certain extent, structured around them. Thus, to give just one well-known example, these functions are formalised and articulated in Mintzberg's organisational configurations (1982). Mintzberg makes a distinction between techno-structure functions (management, finance, logistics, information technology, training, etc.), support functions (human resources, marketing, legal, etc.) and infrastructure functions (maintenance, security, catering, etc.).

As far as the information and telecommunication function and the corresponding technologies are concerned, it is important to point out that, while they may constitute one of the objects (F) of the interactional innovation model (particularly when information technology service providers are brought into the equation), they can also play a fundamental role in the working of the model. Thus they can come into play in various other areas:

1   in linking the various stages of the innovation process (I), (i.e. markets, design and production);
2   in determining the degree of co-production in innovation by facilitating contacts between partners (C);
3   in defining and selecting modes of knowledge processing (M). These technologies affect the processes of producing, processing and diffusing knowledge to the extent that they help to modify the tradability, transportability, divisibility, separability and appropriability of the information (Preissl, 1995; Boland *et al.*, 1996; Antonelli, 1999; Cohendet *et al.*, 1999). They constitute a memory in which some of the experience and knowledge derived from repeated service provision can be accumulated. Some of these technologies, particularly multimedia technologies, allow even tacit or procedural knowledge to be transferred to some extent (Noteboom, 1999).

It is this functional approach, which is generally absent from economic analyses but often seen in management sciences, which will be emphasised here, for various reasons.

- It allows innovation to be considered in all its variety: to take account of legal and fiscal, information technology, financial, marketing, strategic, logistical innovation, etc., without neglecting those objects of innovation located further downstream that are more frequently examined, i.e. product and process in the strict sense. Innovation can affect any of these objects or different combinations of them. Indeed, an innovation (innovation project) often brings several functions or specialisms into play. Thus legal and information technology functions can be seen to be operating simultaneously when, for example, a new software application is designed, and specific means of protection are considered.
- It sheds light on the black box of 'organisational innovations' and process innovations, if it is accepted that they are defined, as noted above, by simultaneous changes in various functions of the firm: human resources, information technology, marketing, communications, etc.
- The various internal functions of the firm can be compared with the corresponding external service providers (the legal function with legal consultancy, marketing function with marketing and market research consultancy, etc.) in order to conduct a more detailed examination of innovation processes. These internal and external functions belong to common disciplinary fields and share a common stock of knowledge.

Stages of the innovation process

In order to understand innovation phenomena, it is not enough to analyse the object of innovation (product, process, or, further upstream, the different functions considered autonomously or from the point of view of their relationship with the new product or process). It is also necessary to examine the process of innovation itself. Indeed, innovation, just like services themselves, is not an outcome. It is an interactive process comprising various activities, which, for simplicity's sake, we will limit to the following (bearing in mind that one preliminary stage is the realisation of the possibility or need for innovation).[6]

- *The gathering of information and ideas on a problem.* The information gathered and transmitted is not limited to scientific and technical information. It may be any kind of information about any of the functions envisaged. This extended definition of information marks a significant break with the Schumpeterian models (Marks I and II), which are often described as 'science-push models'. They are so-called because innovation is very much determined by scientific and technical information and knowledge which are completely exogenous in model I and partly endogenised in model II. Taking the client firm as a reference, the information gathered may be internal or external (i.e. it may relate to the firm's internal or external environment). Information may be collected formally or informally. The gathering of external information, for example, can be formalised in monitoring functions that focus not only on technological and commercial development but also on the whole range of functions in the firm that may be the objects of innovation. The gathering of internal information by members of the firm or by (or in collaboration with) external service providers can also be associated with the activity of problem formulation (diagnostics) – which may in itself be a source of innovation.

- *Research (basic or applied)* in the usual sense of the creation of new knowledge (by combining various stocks of old knowledge). In our view, however, this activity can involve both the exact sciences and the social and human sciences (law, sociology, psychology, economics, management techniques, financial techniques, etc.).
- *Conception and development (C&D)*, i.e., the transformation of the various ideas gathered or produced into a solution to a problem. This activity also includes the test and experimentation phases.
- *Production of the solution* which, in services, is generally inseparable from its commercialisation in that both are co-produced, i.e. the client participates in the production process.
- *Marketing of the solution.* This may take place externally, and in that case would involve the diffusion of the innovation. This phase might also include the establishment, downstream of the process, of various mechanisms (legal or otherwise) intended to protect the innovation. However, this marketing may also be internal. In this case, the innovation would be 'sold' to 'internal clients' of the innovating organisation. This in fact would be a case of 'pseudo-marketing' since, with the odd exception, there is no real organised market. In other words, this phase brings into play all those mechanisms that play a part in the internal implementation or simply in the introduction of an innovation (training, learning mechanisms, etc.) and which are intended to create the conditions under which the innovation can adapt to its new environment, i.e. become 'localised' and 'contextualised'.

As is suggested by the 'chain link' model developed by Kline and Rosenberg (1986), these various activities are not successive phases in a rigid and linear process. Rather, they should be considered as a set of tasks that can be carried out in the course of an innovation process. Indeed, the gap between this perception and a standard linear vision of the innovation process can be gauged by the following observations.

1   The various activities mentioned above concern not only technological innovation in the traditional sense, but all kinds of innovations. For example, a problem within the legal function may well bring some of these activities into play. This is because the search for solutions may include an information and idea-gathering phase, more conceptual analytical phases that resemble genuine R&D (such tasks are,

moreover, entrusted to legal experts, some of whom hold a Ph.D.) and require strategies for appropriation and protection.

2   An innovation process may involve all these activities, or may be briefer and limited to only some of them.

3   The activities defined above may follow on from each other (sequential process), but usually they overlap and some may take place simultaneously.

4   These activities may be formalised (institutionalised) in a 'script' of the innovation process, but are more generally informal and tacit.

5   The service production process may be a stage (or activity) in the innovation process. In some cases, however, these two processes may also merge into one. Indeed, although innovations usually have a certain degree of exteriority in relation to the actors (that is they are clearly identified as innovation projects), there are also innovations which are not planned, which emerge from the process of producing the service and which are recognised as such only after the event. This is the case with what we have termed 'ad hoc innovation'.

The set (I) of elements that make up the innovation process can be combined with a typology of the various types of knowledge brought into play. Faulkner and Senker (1995) have drawn up such a typology that identifies five relatively self-explanatory groups of knowledge brought into play in industrial innovation:

1   Knowledge related to the natural world (scientific and engineering theory, properties of materials)

2   Knowledge related to design practice

3   Knowledge related to experimental R&D

4   Knowledge related to the final product

5   Knowledge related to knowledge (the ease with which the knowledge required to solve a given problem can be located).

This typology of knowledge can be applied to our model provided it is adapted, particularly by extending heading 1 above to include knowledge linked to the social world (social and human sciences). Furthermore, it must take into account the consequences of this extension for the other types of knowledge. This typology will not be introduced directly into our model, since what concerns us above all is not the nature of this knowledge but the ways in which that knowledge is produced and processed.

Consultancy, the client and the degree of co-production

Consultancy itself is a problem-solving activity, as is borne out, for example, by the traditional definition formulated by Greiner and Metzger (1983): 'services provided to organisations by specially trained and qualified persons who assist the organisation, objectively and independently, in identifying and analysing problems, recommend solutions to these problems and help, when called upon, to implement those solutions'.

There is thus some common ground between consultancy and innovation. But not every consultancy service can be considered an innovation activity and the boundary between routine consultancy and innovation must be clearly established.

Furthermore, our interactional innovation model is not limited to the particular category of consultants generally called 'innovation consultants', a term used to denote specialised activities permanently directed towards a clearly defined innovation (a client's innovation project). For example, 'innovation consultancy' in the strict sense would include the work of a patent consultant but not that of a specialist tax lawyer (although the latter plays, as we know, an active role in devising a new insurance contract). Our model goes further in that it seeks to take into account all consultants and consultancy services that may play a part in a client's innovation, whether the innovation exists a priori or emerges from the process of service provision.

Nor is this model limited to services provided by professionals and firms belonging to the consultancy industry as defined by accounting classifications. It encompasses all knowledge-intensive business services, as well as the services provided by R&D laboratories, universities, public intermediation agencies, financial and insurance companies and, finally, inter-firm cooperation in matters of innovation. It can easily be generalised to include an organisation's internal service providers and consultants. Thus it is a model that can apply to numerous situations involving intermediation. 'Consultancy' can be brought to bear on all the functions of a firm. Although space precludes any outline of them here, numerous typologies exist of the whole range of 'problems' for which consultancy activities may provide a possibly innovative solution (legal, strategic, communications consultancy, etc.).

Consultants may act alone or in collaboration or competition with others in the different phases of the innovation process defined above. Thus the aim of their actions may be information or idea gathering, problem formulation (diagnostics), the design, development, testing, production and implementation of a solution, the management of innovation projects, protection and training measures, etc.

Furthermore, the part played by consultants and their clients in these various tasks can take different forms, determining the extent to which the innovation is co-produced. The level of co-production can be zero or very low (e.g. a simple subcontracting situation in which the service provider has a relatively clearly defined brief involving little interaction with the organisation's internal experts). However, it is usually high, which means that the intensity of work at the interface is also high, with the innovation being produced by a partnership between internal experts and external service providers (Leonard-Barton, 1995; Gadrey and Gallouj, 1998). The notion of co-production in the service economy is echoed in the concepts of partnership, collaboration, and learning by using or interacting (Von Hippel, 1976).

The service provider's role in producing and processing knowledge

Another way of looking at the role of consultants in the innovation process, and one which is important for the present argument, is to examine the relationship to information and knowledge (whether embodied or not). This relationship involves the various cognitive procedures used by service providers in the innovation process or the various modes of producing and processing knowledge. Use of this type of service provider (even for routine services unrelated to any innovation project) is justified above all by the greater volume and quality of knowledge, whether static or dynamic (i.e. innovative capacity), that such a provider can bring to bear. In order to take account of this competitive advantage in terms of knowledge, Noteboom (1992, 1999) introduces the notion of 'external economies of cognitive scope', which gives pride of place to a dynamic approach to transaction cost theory. These economies result from the differences or complementary relationships in the cognitive profiles of partner organisations. These differences or complementary relationships reduce the risk of 'missing' certain interesting solutions. In other words, the use of an outside service provider cannot be justified simply on the basis of specialisation and economies of scale but in terms of a more dynamic argument based on economies of variety. Moreover, it would seem that uncertainty does not lead to services being provided internally, as standard transaction cost theory would suggest, but rather to the use of external providers (Noteboom, 1992).

The literature provides some interesting analyses of knowledge production and processing issues. Thus Nonaka (1994) and Nonaka and Takeuchi (1995) put forward a potentially very rich typology of the various modes of knowledge conversion, that is of the 'social interaction between tacit knowledge and explicit knowledge'. These authors identify four modes of knowledge conversion: (1) *socialisation* (from tacit knowledge to tacit knowledge), (2) *combination* (from explicit knowledge to explicit knowledge), (3) *externalisation* (from tacit knowledge to explicit knowledge), (4) *internalisation* (from explicit knowledge to tacit knowledge). Noteboom (1999), for his part, draws inspiration from Piaget (1970) in order to describe a learning cycle involving five principles: (1) *generalisation*, i.e. the application of a practice to new but related contexts, (2) *differentiation*, i.e. adaptation to the local context, (3) *reciprocation*, i.e. 'the exchange of elements from different parallel practices, in a given context', (4) *novel combinations or accommodation*, i.e. the combination of elements derived from different practices which, together, constitute a new practice, and (5) *consolidation*, i.e. the transformation of a new practice produced through the process of accommodation into the 'dominant design'.

The modes of knowledge production and processing adopted for our model have certain points in common with those listed above, although they are by no means identical. For example, unlike Nonaka and Takeuchi's typology, our approach stresses not so much the nature of knowledge (codified or tacit) as the technical nature of the mode of processing brought into play (simple transfer, combination, dissociation, adaptation . . .). In this way, our model reflects more closely the actual modes of operation adopted by service providers.

A consultant's role may, indeed, be limited to the *algorithmic or linear transfer* of information and knowledge[7] that may be embodied in humans or in technical systems (technological transfer). This mode of activity seems to occupy an important position in certain areas, such as marketing research, for example. This concept of consultancy is derived from a standard perception of technological information as a quasi-public good, i.e. as a non-excludable, non-appropriable or non-rival good, which can be transferred easily and at low cost (Arrow, 1969). However, an activity limited to this algorithmic (or linear) transfer function seems to contradict the very existence of consultancy activities. Indeed, the fact that these activities exist, and at a not inconsiderable cost, proves that information is not accessible at low cost, omnipresent and easily transferable. Indeed, other relationships to information and knowledge (or more precisely the transformation of information into knowledge)[8] can be brought to light, and these processes of knowledge production and transformation fall within the Schumpeterian and neo-Schumpeterian tradition. We may distinguish:

- *(Creative) combination*: the aim here is to create links, not only between information and knowledge, but also between people or organisations. Various activities can be described through this mechanism:
    - R&D, which consists of creating new knowledge by (creatively) combining old stocks of knowledge;
    - searching for partners, which involves combining organisations;
    - the role of 'marriage broker' – to use the term coined by Bessant and Rush (1995) – which also involves combining organisations but where a consultant is responsible for 'recruiting' organisations and managing the interaction between them.
- *Knowledge localisation, contextualisation or 'customisation'*; i.e. the transformation of standard (generic) information into tacit, idiosyncratic, cumulative and path-dependent knowledge adapted to the client's particular situation, making it appropriable by the client but not readily transferable to others. 'Localisation' thus transforms generic information or knowledge into quasi-private goods (Antonelli, 1995; Atkinson and Stigliz, 1969; Petit, 1998).

Table 3.1 Different opposite modes of producing and processing knowledge

Modes of knowledge processing and production	Opposite modes
Combination, association	Dissociation
Learning	Unlearning
Localisation, contextualisation, customisation, creative adaptation	Formalisation

- *Knowledge formalisation* which consists, conversely, of making knowledge more objective, less tacit and more transferable, i.e. generic, by constructing or integrating it into a social context. The formalisation of a problem and establishment of a diagnosis may also fall within the scope of this mechanism which, in this respect, has similarities with the '*translation*' mechanism of network sociology (Callon, 1986), i.e. the 'enlistment' of real or potential actors to help in defining a problem. Indeed, the definition of the nature of the problem is not 'self-evident' but largely socially constructed.
- *Learning.* This innovation mechanism must be considered in all three of its aspects: 'learning oneself', 'teaching others', i.e. clients, through formalised processes (training) or non-formalised processes (learning by interacting), as well as 'teaching others to learn', i.e. maintaining and improving what Cohen and Levinthal (1989) call firms' absorptive capacities.

It is clear that the various mechanisms through which the 'consultant' shapes and fashions 'knowledge' must be considered both positively and negatively (Table 3.1). Thus, the opposite of combination is dissociation, i.e. the destruction of links between knowledge, organisations, etc., that of learning is unlearning, i.e. the destruction of obsolete knowledge while that of localisation (i.e. knowledge contextualisation) is formalisation (i.e. the codification of knowledge, which creates the conditions under which it can be transferred).

The positive or negative nature of a particular process does not imply any value judgement: unlearning, for example, is not a disease within a firm but rather a necessary mechanism that is often difficult to implement (Hedberg, 1981; Imai *et al.*, 1986; Johnson, 1992) but provides firms with a means of escaping from 'competency traps' (Levitt and March, 1988) or from cognitive or technological lock-in situations (David, 1985; Arthur, 1989) in order to implement new learning trajectories. Similarly, dissociation, as opposed to association (or combination) creates knowledge. It even constitutes a model of innovation production that is widespread in services (Gallouj and Weinstein, 1997).

Each of the facets (positive or negative) of a given process can manifest itself independently, but they can also precede or succeed each other. Thus, learning can follow unlearning and a network can be reconstructed following the break-up of the previous configuration. Some of the modes of processing knowledge may be stages in a sequential process: for example, knowledge may first be formalised so that it can subsequently be combined more easily and finally transferred.

Comparison of our typology of modes of knowledge processing with those of Nonaka and Takeuchi (1995) and Noteboom (1999) gives rise to the following observations:

1   Formalisation can be considered to be synonymous with externalisation. It covers the two principles of consolidation and generalisation.
2   Localisation corresponds to the principles of internalisation and differentiation.
3   The notion of combination is fairly close to that of accommodation. On the other hand, it is more extensive than Nonaka and Takeuchi's concept of combination. It also covers what those authors term socialisation, since tacit knowledge can also be combined. In their typology, the conversion process that is the 'opposite' of combination is not dissociation but socialisation.
4   The process that we have entitled linear or algorithmic transfer falls within the scope of Nonaka and Takeuchi's notion of combination, since it involves the transfer of codified knowledge.

In conclusion, the various modes of producing and processing expertise are not specific to situations in which an external service provider has a role. Any production and processing of knowledge, whoever the actors may be, can be considered in the above terms. Our interactional model of innovation does not, therefore, apply solely to consultants or, more generally, to external service providers but can also describe the activity of any internal researcher or innovator as well.

The innovation model as an interactional process

Innovation relates to a function. It presupposes a certain mode of relationship with the client and a certain approach to the processing of knowledge. In general terms, the interactional model of innovation can be formalised by articulating the following four elements (see Figure 3.1):

1   the components (stages or activities) of the innovation process in which the service provider may play a part (I);
2   the functions (F) in the client firm that are the object of innovation activity. This may be any function (or combination of functions) in the firm, and not only those that play a part, downstream, in the design of a new product or process;
3   the degree of involvement of the service provider (and client) in the innovation, i.e. the degree to which the innovation is co-produced (C);
4   the different modalities (M) of knowledge production and processing which may be implemented by the service provider.

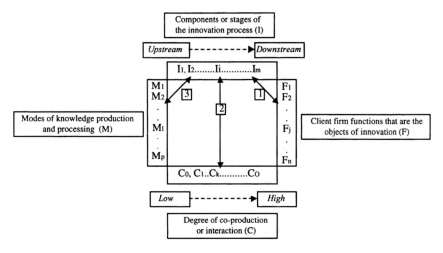

Figure 3.1 The interactional model of innovation

Description of the model

If, first, the relationships between vector **I** (activities or stages in the innovation process) and the other components of the model are examined, the
following observations can be made:

- The multiple combinations possible between the various stages or activities in the innovation process and the various functions in the firm
  (relationship 1 in Figure 3.1: $\{Ii\} \leftrightarrow \{Fj\}$) represent the functions or
  groups of functions affected by the innovation (or by certain stages or
  components of the innovation).

Thus innovation (the process in its entirety or some of its stages) can relate
to any function or group of functions Fj independently of any product or
process innovation, and not only to the 'functions' corresponding to the
new 'product' or 'process'.

If it is assumed, for example, that I1 represents the gathering of information and ideas, I3 the development stage (or, more precisely, the implementation of a test), F1 the legal function and F5 the insurance settlement
function, then:

the relationship I1 $\leftrightarrow$ F1 denotes 'the search for information and ideas
for a problem relating to the legal function';
the relationship I3 $\leftrightarrow$ F5 means that a test is carried out in the insurance settlement function.

- The combinations {Ii} ↔ {Ck} (relationship 2) represent the degree of co-production not of the service but of the innovation (or some of its stages), although, as we have already pointed out, the two processes are inseparable from each other in certain cases, in that the innovation can emerge from the service provision process. The degree of co-production can vary from one task to another (e.g. it can be low for Ii, information gathering, but very high for I3, development).
- Combinations {Ii} ↔ {M1} (relationship 3) represent the various forms taken by the service provider's intellectual input into the various stages of the client's innovation process. Thus, I1 ↔ M1 symbolises the simple mechanical transfer of information and ideas and I3 ↔ M3 the production of new knowledge (R&D activity based on the creative combination of old stocks of knowledge).

Different configurations of the model

If we confine ourselves for the moment to the role played by a single consultant, the interactional model of innovation offers a considerable number of combinations {[Ii], [Fj], [Ck], [M1]}. Thus, the multiple combinations of the different vectors, the components of which can themselves also be combined, make it possible to conceive of a very large number of 'co-produced' innovation spaces.

This general model cannot be reduced to a few stable, clearly identifiable configurations. However, a standard configuration of the model can be contrasted with a (generic) configuration which will be termed 'evolutionary'.

The standard configuration is easy to identify. It constructs a clearly delimited single innovation space (see Figure 3.2). There is zero co-production (C0) since there is little if any interaction between the service provider and the client. This is in fact a subcontracting or jobbing relationship. The mode of processing this technological knowledge is simple mechanical transfer (M1).

For sets {Ii} and {Fj}, however, the choices are numerous:

- the functions Fj in question are those required to produce new goods or services, if the object of the innovation is that good or service. More generally, the object of the innovation could be any function or set of functions, independent of any goods or services;
- as far as the innovation process is concerned, once again, different 'phases' Ii or groups of 'phases' may be involved.

These various operations relate to an innovation process considered to be linear, i.e. one in which the various activities are independent of each other: there is no interaction or feedback.

In other words, in the standard configuration, the model is reduced to the transfer of codified (technological) information, feeding into one or other

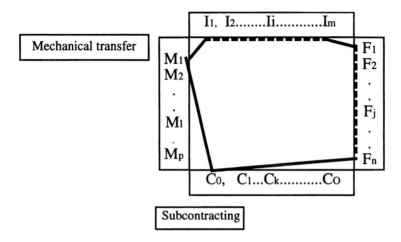

Figure 3.2 Standard configuration of the interactional (or assisted) model of innovation

or all the stages of the innovation process, without any co-production what-soever, i.e. without any interaction with the client. A typical example of this standard configuration is the transfer of technology (turn-key projects) to developing countries, in which some consultancy and engineering firms have involved themselves and which has sometimes resulted in resounding failure. However, it is important to remember that this form of transfer can also involve non-embodied technologies, management systems, methods, cognitive families and functions.

However, the standard configuration is only an extreme case of our model, which allows many other configurations or 'states' to be envisaged. Evolutionary theory is concerned with systems that have a wealth of inter-actions (Coriat and Weinstein, 1995) and variety (Saviotti, 1996; Metcalfe and Gibbons, 1989). It introduces multiple modes of knowledge processing. At the same time, it puts forward a concept of innovation as the resolution of selected problems (Dosi, 1982) and a notion of the firm not as informa-tion but as knowledge processor (Fransman, 1994). Thus all co-produced innovation spaces where transfer is not limited to its simple mechanical form can be considered as evolutionary configurations.[9] Irrespective of its 'state' at any given moment, the model always presupposes that the knowledge tapped or produced is transferred (to the client). However, this transfer is accompanied by a more or less complex series of manipulations and pro-cessing operations (contextualisation, formalisation, association, etc.).

Since it is impossible to reduce the immense diversity of evolutionary configurations that the interactional model of innovation can take to a few typical cases, we will limit ourselves to a few examples which we have observed in surveys: Gadrey *et al.*, 1993; Gadrey and Gallouj, 1994; Djellal *et al.*, 1998. In order to simplify the graphic representations, we will not

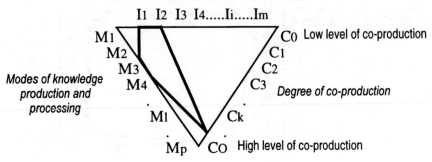

Figure 3.3 Ad hoc innovation configuration

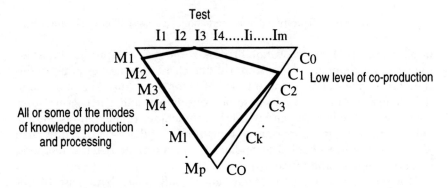

Figure 3.4 Consultant-assisted test model

take component (F) of the model into account. In other words, in the various configurations presented here, the function or functions that are the object of innovation are considered as given.

1 The co-produced innovation space delimited by the set {I1, I2, CO, M3, M4} (see Figure 3.3) can denote, for example, the co-production of a solution to a problem relating to the legal function, which we have elsewhere called ad hoc innovation (Gallouj and Gallouj, 1996). I1 and I2 represent, respectively, the gathering of information and ideas and an analytical and conceptual activity comparable to R&D, CO, a high degree of co-production and M3 and M4 processes of knowledge contextualisation and combination. In this configuration of the model, innovation cannot be envisaged independently of the process of service provision and the actors involved. It is constructed as the service itself is provided.

2 The role of a consultant in the drawing up of an insurance contract (and, more precisely, in the test phase of this new contract) is illustrated in Figure 3.4. This is a frequent configuration in services, one that might

be called the 'consultant-assisted test' configuration. It is delimited by sets {I3, C1} and {M1} in which I3 represents the test phase, C1 a degree of co-production or interaction greater than zero and {M1} the whole range of knowledge-processing methods;

3 The externalisation of R&D, i.e. the use of external laboratories (public or private), universities or grant holders, also falls within the scope of our interactional model of innovation. Occasionally, it can operate according to the standard algorithmic configuration (subcontracting and mechanical transfer), but the interactive mode is more usual. The stages of the innovation process involved here are the research and development phases (I2 and I3) and the knowledge processing methods used can be any of {M1}. One of the variants of the model of R&D externalisation is illustrated in Figure 3.5.

Whatever the example considered, the use a given firm makes of an external service provider (in this case, as part of the innovation production process, but this can also apply to routine services) cannot be considered independently of the question of the importance of the role played by the corresponding 'internal provider'. As far as R&D activity is concerned, many theoretical and empirical studies adopting a variety of different approaches (game theory, evolutionary theory) have sought to ascertain whether the externalisation of R&D and the execution of those activities within the firm are substitutable or complementary (Kleinknecht and Reijnen, 1992; Colombo and Garrone, 1996). Similar questions have been asked of the use of knowledge-based services (Gadrey and Gallouj, 1998; Quinn, 1999). These empirical studies tend to find a degree of complementarity between internal and external services. Over and above mere technological knowledge, they extend the notion of absorptive capacity (Cohen and Levinthal, 1989) to knowledge relating to all the functions (F) of a firm. Thus performance in the production and processing of knowledge by an external service provider depends on the existence of professionals within the firm who can act as an interface.

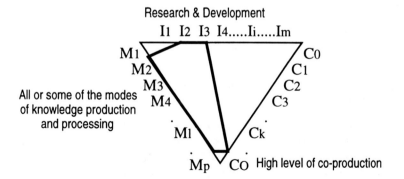

Figure 3.5 A variant of the model of interactional R&D externalisation

Model of innovation assisted by (several) consultants

The model can be made more complex by having several consultants contribute to an innovation. Once again, a large number of configurations are conceivable. For the purposes of illustration, we will limit ourselves to the following four, only the first two of which are represented in Figure 3.6.

This illustrates the configuration in which a single innovation or stage in the innovation process depends on several consultants (operating competitively or cooperatively, and with varying degrees of interaction) to solve a problem relating to a single function. The participation spaces of the different consultants overlap to varying degrees, depending on the type of interaction implemented and the knowledge-processing method used. In configuration 1 in Figure 3.6, the types of interaction and knowledge-processing methods are identical for the different consultants; the configuration in which, for a problem relating to a given function (Fj), or group of functions, which may be autonomous or linked to a wide-ranging product or process innovation project, different consultants are required for the different stages in the innovation process. For an important technical process innovation project, for example, a firm might use – at the same time or at different moments in time – marketing and monitoring specialists to gather information, knowledge and ideas, one or more university laboratories to solve technical R&D problems, a project management consultant to manage the innovation project and integrate the various participants; a patent consultant, etc. (see configuration 2 in Figure 3.6); the configuration in which a different consultant is mobilised for each function. This configuration differs from the one in which most of the problems relating to the different functions can be tackled by the same multi-specialist network consultancy firm; the configuration in which a consultant calls upon another consultant (e.g. an academic) for assistance. This configuration could be termed 'consultant-assisted consultancy'.

Theoretical implications of the interactional innovation model

The interactional innovation model raises a number of interesting theoretical issues which will now be examined.

Beyond the micro-economic (service) relationship: system and network

This chapter has thus far focused on micro-economic relationships. But our model can be fully understood only if it is relocated in a wider system of meso or even macro-economic relationships which condition the way it operates.

The meso or macro-economic effects of such a model are also linked to the roles played by consultants and, more generally, by knowledge-

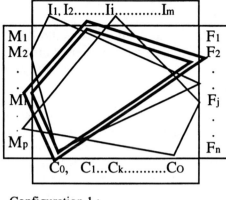

Configuration 1 : ▬▬
Configuration 2 : ──

Figure 3.6 Various configurations of the model of innovation assisted by (several) consultants

intensive business services in innovation systems and (extended) techno-economic networks.

The evolutionary approach to the economics of innovation and the sociology of innovation and networks are closely akin to each other, in that they both take interaction between agents as their starting point. Thus consultants are both elements (nodal points) in local, regional and national innovation systems (Lundvall, 1985, 1988) or techno-economic networks (Callon, 1991), and above all, vectors instrumental in the establishment of relationships between agents. They create links – what Granovetter (1973) calls 'bridges'. They maintain 'strong ties' with their clients[10] but their principal value lies in their ability to exploit to their clients' advantage more tenuous links, links weakened by the various (additive) forms of 'distance' *vis-à-vis* the clients in question. This 'distance' may be temporal (we came across a similar problem a while ago), sectoral or functional (we came across a similar problem in a completely different sphere), geographical (we encountered this problem in a different country) or even symbolic (prestigious academic networks of *grandes écoles* in France).

Thus consultants (and in many case internal experts as well) can be seen as entities where three types of network intersect (Decoster and Matteaccioli, 1991):

- *international business networks*, which are the media for the transmission of information from other geographical spaces. These networks can take different forms (networks of correspondents, subsidiaries, etc.). When these 'correspondents' support the consultant or bring specialist experience to bear, we could be dealing with a reflexive model that constitutes a particular variant of the general model, i.e. the 'consultant-assisted consultant model' already referred to above;

- *institutional diffusion networks*: relationships with public authorities;
- *prestigious academic networks*: relationships between consultants and the most famous universities or the *grandes écoles* (in France) and between the graduates of those institutions.

In (extended) techno-economic networks, consultants can be actors not only in transfer zones but also in central zones (scientific, technical, market poles), to use the distinction made by Callon (1991). Thus, internal and external consultants can be classified according to their area or expertise, whether it be science, technology or the market, or the transition from one area to another. Consultants also play a role in the morphology and dynamics of the techno-economic network. Thus, they are able to play a part in reconfiguring or extending the network by adding other actors, whether in a real or virtual way (i.e. by mobilising 'weak ties' in their production function). They can also facilitate the process of 'translation', in the sense of the term used in the sociology of networks (Callon, 1986), and contribute to network convergence (i.e. to the free flow of information, knowledge and solutions within the network).

Interactional innovation and the convergence of innovation systems

This chapter advances a view of the service provider and the service relationship as an expression of the Schumpeterian spirit of enterprise, made available to clients or activated by them. In other words, 'consultants' are regarded as 'extra cognitive memory' for firms which are now defined not as production functions but as 'learning organisations'. Our hypothesis is that they maintain economic diversity, what might be called, by analogy with 'biodiversity', 'eco-diversity' or 'socio-eco-diversity', and that they activate the processes of knowledge production and organisational learning (at least if their brief is successful).

However, just as Schumpeter feared that the endogenisation of the entrepreneurial function in R&D departments would lead to bureaucratic inertia and to the stifling of the engine of capitalism, it is possible to advance a view of consultants as contributors to the convergence and increasing irreversibility of innovation systems, i.e. to the eventual reduction of the degree of variety in those systems. In this scenario, consultants would have a damaging effect since their actions would lead to a decrease in eco-diversity; as examplified by the extinction of economic species, as it were. In the long term, economic systems would become increasingly uniform through the application of the same technical systems, types of organisation, management systems, as exemplified in the international diffusion of accounting, financial, recruitment and management norms, 'lean production', etc. This 'cognitive' convergence is said to be taking place already, through the intermediary of international consultancy networks of Anglo-Saxon origin. In France, some sociologists (Henry, 1992) attribute this convergence to endogamous economic behaviour generated by the uniquely French system of producing and reproducing elites through the *grandes écoles*. The risk of network lock-in is thus the principal 'weakness of the strong ties' thus created, to use A. Grabher's terms (1993). If it is assumed that economic variety is preferable to uniformity, a number of arguments can be put forward that may help counter this risk of convergence (as a dominant trend):

1 The first argument relates to the 'localised' or 'contextualised' nature of the knowledge, know-how and innovations co-produced and diffused by consultants. As has already been stated, this knowledge is often 'reconstructed', since it tends to lose its shape on contact with different firms, defined as multidimensional economic spaces (Noteboom, 1992; Antonelli, 1986).[11] As it comes into contact with the various points in these spaces, it is subject to displacements and deflections. Furthermore, these spaces, which might more appropriately be called multidimensional socio-economic spaces, are not static but fundamentally dynamic. They change over time, which is a source of eco-diversity. This accords with several theories in spatial economics, according to which the spatial diffusion of a dominant mode of organisation on a worldwide scale does not homogenise the space. On the contrary, diversification occurs as this mode of organisation comes into contact with historical, cultural, economic, etc. specificities and with the particular dynamics of innovation at work. Spatial diffusion gives rise to as many different innovation trajectories as there are innovative 'environments' ('*milieux*') (Aydalot, 1986; Gordon, 1990; Decoster and Matteaccioli, 1991). This same idea appears in recent studies from the French regulation school. That 'industrial models travel in search of favourable spaces and are transformed by them' is one of the main conclusions of '*Le monde qui va changer la machine*' (*The world that will change the machine*), (Boyer and Freyssenet, forthcoming).

2 As is suggested by some studies, particularly in the area of organisational sciences (Eisenberg, 1984; Fiol, 1996), organisational or collective learning is a process based on a paradoxical mechanism. Eisenberg (1984) calls this 'unified diversity', in which a certain balance is achieved between two apparently contradictory mechanisms, namely a diversity of interpretations, on the one hand, and unanimity (i.e. consensus building), on the other.

3 The power of clients (especially the largest ones) and their cognitive qualities must not be underestimated. The risk of convergence is a function of the balance of power and mutual influence between the client and the service provider. A subcontracting relationship can more easily lead to convergence in the long term, i.e. the standardisation of products, processes and methods, etc. There can be a great (Procrustean) temptation to make all problems fit the existing solutions.

However, a more interactive relationship, which takes greater account of clients' specificities and their internal and external environment, will be a source of diversity and new solutions, since the situations are themselves always new. Under pressure from clients (i.e. when they have sufficient knowledge to make their 'voice' heard, to use Hirschman's term (Hirschman, 1972)), consultants make efforts to 'localise' their information and knowledge so that they adapt them to the 'need' at the same time as they 'reconstruct' that need. Co-production enables the client to leave his specific personal mark on the knowledge. It is a factor in syncretism and hybridisation, both mechanisms that encourage eco-diversity. Furthermore, the client in an innovation project can call on several consultants and have them work cooperatively or competitively in order to ensure a measure of diversity.

There is undoubtedly a greater risk of convergence in the design and introduction of technical artefacts than in more malleable, intangible solutions. Nevertheless, even those consultants most involved in handling technical artefacts (e.g. information and communication technology consultants) are guided less by deterministic technological trajectories than by socio-technical trajectories which establish a balance between the technical, organisational and social dimensions (Djellal, 1995). In other words, since the physical equipment cannot be separated either from the software or from the social environment, the risk of convergence inherent in technical systems is reduced.

4 The consultants' own capacity for innovation must not be underestimated either. This is a source of diversity and of differentiation from competitors and reduces the risk of convergence. Even if consultants are instruments of the 'laws of imitation', it should not be forgotten that the universality of these laws (as defined by Gabriel de Tarde, 1890) can take a paradoxical turn: to imitate is also to differentiate or to do the opposite of the object of imitation. To quote the very words used by Tarde (1890, p. 11 of the preface to the second edition), 'there are in fact two ways of imitating: to do exactly the same as one's model, or to do exactly the opposite'. In other words, benchmarking does not exist, or only as a 'mobilising slogan' within firms, since 'it is not innovation that changes the world but the world that changes innovation'.[12] In consultancy, moreover, the existence of 'networks' on different levels creates 'distortions' in the transfer process that alter the nature of the knowledge being transferred. The more extensive the network, the lower the risk of convergence becomes. Indeed, there is loss, selection and distortion of knowledge during its entire progression through the network's internal spaces. And as an external source of knowledge, it is in an outside service provider's interest to maintain a certain 'cognitive distance' from his client if he wishes to benefit from 'external economies of cognitive scope' (Noteboom, 1992).

4 Unlike the Schumpeter Mark I and II models, which are 'science or technology push' models, interactional models of innovation offer a balance between the science-push and demand-pull approaches. Indeed, clients

and their needs (represented by the support functions [Fj] or 'objects' of the problem to be solved and by the intensity [Ck] of the client/consultant relationship) are a central element of the model. It must also be noted that this model enriches the content of the scientific determinant in that it also takes social and human sciences into account. This co-determination by science (understood in a wider sense) and demand is also a source of variety which limits the risks of convergence.

Consultants and the risks of technological and cognitive lock-in

This problem is closely linked to that of convergence discussed above. Either voluntarily (to secure customer loyalty) or involuntarily (because of expertise differentials), consultants and the consultant-assisted model of innovation can lead to technological or cognitive lock-in, i.e. cognitive dependence and complementarity phenomena that are disadvantageous to the client or system. The notion of lock-in is understood here in a wide sense. Thus, for example, a legal consultant could put together a legal and fiscal arrangement of such complexity that nobody else would be able to find their way around the maze. This problem is more familiar, if not more common, when physical technical systems or software are involved (e.g. computer consultancy). Methodological lock-in (in the case of intangible technologies) is also common.

However, it should be noted that this lock-in phenomenon may, conversely, work in favour of the client himself, when it is the consultant who is subordinate to his client, i.e. when the balance of cognitive and technological power is tilted towards the latter. In many situations, moreover, the consultant is required, conversely, to help the company to extricate itself from its own 'competency traps' or to unlock its own internal lock-ins.

The appropriation of co-produced innovation

A service provider's participation in the production of a client's innovation raises serious problems of appropriation. The difficulty of appropriating innovations makes itself felt both positively and normatively. It is technically difficult, even impossible for the service provider or the client to protect the innovation, which is very often a service innovation (relating to the service functions of manufacturing or service firms). This is typical of intangible and relational activities.

This positive or technical difficulty (absence of legal methods of protection) is compounded by another, which might be described as a normative difficulty. This is because it concerns the sharing of ownership rights over a co-produced innovation, in respect not of the technical modalities but rather of the designation of the (legitimate) owner of the innovation. Do the co-produced innovation and the knowledge stemming from the co-production process belong to the client or to the service provider? This problem is all the more difficult since the innovation process described by the interactional innovation model does not necessarily equate to an identified project. It may be 'non-programmed' and may emerge as a routine service is being delivered.

Different arguments can be put forward:

- In the standard configuration of the model, the problem of the appropriation regime does not arise, as the information and knowledge do not belong to the innovator. As public goods, they are non-rival and non-excludable, i.e. they cannot be appropriated by the innovator.
- More generally, though, knowledge is shaped in such a way as to be adapted to the 'topology of a multidimensional socio-economic space'. This is localised knowledge that lies outside the mechanical universe of information and knowledge as (non-excludable and non-rival) public goods that are both transparent and transferable. Information and knowledge derived from the interactional model of innovation are adapted to the client's socio-oeconomic problem and thereby become quasi-private goods. They thus acquire an idiosyncratic character, which facilitates their appropriation by the client. The client enjoys a partial monopoly on this knowledge. His competitors cannot copy it, for reasons that are very clearly described in the literature, and in particular the existence of 'interpretative ambiguity' (Fransman, 1994) or 'causal ambiguity' (Lippman and Rumelt, 1982) and the weakness of 'absorptive capacities' (Cohen and Levinthal, 1989). On the other hand, non-localised (raw) information (generic information) is not of this kind: the service provider can make use of it and 'localise' it elsewhere without harming previous clients.
- Given the nature of innovation and the rise to prominence of what we call the recombination or architectural model, i.e. a model of innovation based on the principles of dissociation, association and the systematic re-utilisation of existing 'components' (Henderson and Clark, 1990; Foray, 1993; Gallouj and Weinstein, 1997), the problem or issue increasingly seems to be not the protection of knowledge but rather the mechanisms likely to facilitate its diffusion and therefore its multiple recombinations. Consultants are thus 'spillover' factors that may help to establish a social optimum. If, as Jaffe (1986) shows, a firm's R&D productivity is increased by the R&D of its 'technological neighbours', it is likely that knowledge-intensive business services constitute a vector for this increase in productivity, as well as being one of the contributory factors in the development of this 'technological proximity'.

One interesting question, closely linked to that of appropriation, is worth mentioning here, namely the effect of interactional innovation on company performance. This may well be fertile ground for application of the methodologies used in studies of technological alliances (Hagedoorn and Schakenrad, 1994). However, they will have to be adapted, since technological alliances generally take place between manufacturing rather than service firms, and innovativeness is generally measured in terms of 'patent intensity', which is seldom relevant in the configurations we consider. They also require the construction of a database of cases (which, to the best of our knowledge, does not currently exist).

Is the interactional model of innovation consonant with the notion of 'waves' of creative destruction?

It can be hypothesised that consultancy activities may help both to keep clients in the 'circular flow' (i.e. prevent a firm from disappearing in the turmoil of creative destruction) and to facilitate their exit from it (through innovation). They can keep the client in the 'circular flow' in two different ways: by departing from their own circular flow (i.e. by innovating) or by staying in it (more routine service provision).

Consultancy can also be seen as an activity which artificially keeps (or seeks to keep) moribund firms in the economic circular flow, although Darwinian mechanisms of 'creative destruction' should lead to their disappearance. Thus, it could be said that the consultant-assisted model of innovations runs counter to the Schumpeterian dynamic of waves of creative destruction. In reality, though, there is no contradiction with the Schumpeterian spirit, since it is not possible to speak of an assisted innovation model unless the consultant introduces an innovation or participates in an innovation which prevents the client from disappearing. When the consultant is involved in the provision of routine services, we are not dealing with the assisted innovation model.

Conclusions

This chapter has highlighted an innovation model which extends traditional Schumpeterian models. This model, in which one or more consultants (the term being used in a wide sense) assist their client in a clearly identified or emergent (i.e. non-programmed) innovation, can take numerous configurations depending on the function or functions of the firm which constitute the object of innovation, the stage or stages of the innovation process in question, and the methods of processing and producing knowledge which are considered. The different configurations of this model depend on a number of factors including the consultant's style, the client's style, the nature of the problem, etc.

The standard configuration of the model, in which knowledge processing is limited to mechanical transfer and interaction is minimal, is only an extreme case. The model is primarily evolutionary, rich in low-level or intense interaction in a range of different spaces: temporal, functional, geographical and symbolic.

In sum, the model rests on a broader definition of innovation. It takes account not only of all the heterogeneity, which is often concealed behind the terms innovation and organisational change, but also of the semantic diversity inherent in the various notions of product innovation – new goods, new services (intangible products), new solutions (ad hoc and tailor-made products) – and in the notion of process: technologies, methods, etc.

Although our analysis has focused on the role of services (and particularly knowledge-intensive business services) in their clients' innovation process, our model can be applied to intra-firm service relations. It can also be applied to inter-firm alliances, collaboration and cooperation, irrespective of sector. In each of these cases, the degree of interaction can be greater or smaller, different stages of the innovation process may be involved, several modes of knowledge processing may be brought into play and different functions may be the object of innovation. The model's usefulness, then, seems considerable.

Notes

1 See Petit (1986), Delaunay and Gadrey (1992).
2 On this point, see F. Gallouj (1994b); C. and F. Gallouj (1996); Gallouj and Weinstein, 1997; Sundbo, 1998a and studies from the SI4S project, financed by the European Commission (programme TSER, DG XII). The SI4S project, completed in June 1998, brought together ten European teams from Denmark, France, Germany, Great Britain, Greece, Italy, the Netherlands, Norway and Sweden to work for two years on the theme of innovation in and by services.
3 These less restrictive terms were suggested to us by J. Gadrey.
4 Recent statistical studies confirm this result (see Djellal and Gallouj, 1998).
5 By analogy with terms traditionally used in economic analysis 'capital-intensive' and 'labour-intensive'.
6 For a concrete examination of innovation processes in different service activities (consultancy, banking and insurance, electronic data services) see Gadrey J. *et al.* (1993, 1995).
7 These two terms being, in this case, synonymous.
8 In other words, these two terms are no longer considered to be synonyms, since, in this case, knowledge is information which has been sifted, processed and contextualised by intelligence.
9 Other theoretical approaches give prominence to the interactions and/or diversity of the modes of knowledge processing, in particular systems theory, non-equilibrium thermodynamics, organisation theories and symbolic interactionism, as well as Austrian economics (see Hodgson, 1998) and resource-based theories (Montgomery, 1995).
10 The strength of these ties is partly expressed by the degree of co-production in our model {Ck}.
11 According to this author, the main topological dimensions of these economic spaces are: the scientific space, the product characteristics spaces, the consumption space, the function externalisation space, the inter-industry relations space as defined by input/output tables, the work space, the geographical space, etc.
12 These remarks are borrowed from Robert Boyer, Clersé Seminar, 28 May 1998, Lille.

4 Innovation as a strategic process

Jon Sundbo

Introduction

Innovation has very much been seen as pushed by technology, but often it is not technological but social (e.g. a new organisation or a new service through a changed behaviour from the service personnel). It is often pull-oriented developed due to market possibilities. Very often innovation consists of small changes, woven together in a complex pattern, (which possibly may not even be called incremental innovations). The radical innovations are rare and their effects in form of creating technological trajectories (cf. Dosi 1982) are rare. Technological trajectories may influence the innovation decisions in firms, but they will generally be mixed up with other considerations such as market possibilities, internal resources, or other types of trajectories such as managerial (new ideas of managing and organising), service professional and social (social movements in society, e.g. ecological) (cf. Sundbo and Gallouj 1999).

This conclusion leads us to an approach to innovation which emphasises the market pull and the internal organisational processes within firms to explain innovation. It gives another answer to the questions why, when and how of the innovation process than the push-oriented technology-economic tradition (e.g. Dosi *et al.* 1988, Freeman and Soete 1997, cf. discussion in Sundbo 1998a). Here the answers will be:

- *why innovate*: because the firm is losing to market competition
- *when innovate*: when profit or turnover is stagnating
- *how innovate*: by creating a strategy and having internal organisational processes

The choice of topic of this chapter does not mean that the technology-push approach is not relevant and cannot explain innovation. It can, but not completely. It must be supplemented with explanations which take their point of departure in the market-pull and the innovation process as a complex social process. Development of such explanations have not been very common. This chapter is one attempt among others to formulate such an approach, which must be considered as complementary to the technology-push models.

The understanding developed in this chapter emphasises creativity and interaction. This can also be a supplement to the knowledge oriented models and theories in economics (Machlup 1980), which have a tendency to interpret knowledge in a rather rational statistical manner (cf. OECD 1996): it can be measured and counted. This is one aspect of knowledge, but it can be supplemented with another aspect, which is creativity and knowledge creation as an interactive process (and not something you buy from an R&D supplier).

These aspects of innovation have not been completely neglected. They have been treated – among others – by Burns and Stalker (1961), Kotler (1983), Rumelt (1987), Nyström (1979, 1990), Tidd *et al.* (1997). In relation to these different analysis, this chapter particularly focuses on the management aspects with strategy as the core concept. That can add some new aspects to the general movement towards a new understanding of innovation.

Empirical examples, primarily from the service sector, will be included in the chapter to illustrate some of the points. The examples are mostly taken from a Danish project (service development, internationalisation and development of competencies, SIC 1999). The chapter is also based on other case studies I have undertaken, primarily in the service sector (e.g. Sundbo 1998b).

There is fairly clear evidence for this approach (as well as that of the other authors mentioned) being adequate to explain innovation in services (Gallouj 1994b, Miles *et al.* 1994, Brentani 1989, 1993, Sundbo 1998b, Sundbo and Gallouj 1999). We might take some analyses (e.g. Nyström 1979, Tidd *et al.* 1997) as evidence for it is – at least partly – adequate to explain innovation in manufacturing.

The looking for a new innovation explanation: more Kirzner than Schumpeter

In this chapter I will develop and discuss an understanding of innovation as an organisational process that is guided by external conditions: market possibilities and other pressure from the environment. Strategy is the core concept which holds the analysis together. Strategy implies, in this case, an external aspect: the firm's market position as well as an internal aspect (how to develop the organisation to meet the market demands (cf. Pettigrew 1985, Mintzberg and Waters 1982). Strategy is on one hand rational firm behaviour towards market competition, on the other hand, it is a social and political process within the firm's organisation.

Economic and sociological elements will be combined in the attempt to explain firms' innovative behaviour. The innovation process will be seen as an interactive process in which the social relationships between actors or role players are important. This introduces other rationalities than simply economic ones, and other demands on innovation management than R&D and project management.

The chapter thus contributes to the attempt to develop a new theoretical understanding of innovation by presenting some elements. These will be conceptualised through two notions, strategy and dual organisation. Strategy is the management's reflection over possible developments of the environment. It will also lead to the other core point of the discussion, the internal organisational processes because strategy here is seen as a means by which the manager leads this process. The organisation of the innovation process is considered a dual one, which means that employees and middle managers are creative and take initiatives for innovation activities, but the management also does that. Moreover, the management also controls the free innovation process in the organisation. The strategy functions as an internal incentive and control system by stating some values and goals and being a framework for the innovation activities.

In relation to the overall aim of the book, this chapter discusses how the firm acts as a mediator in relation to the market. Innovation is seen as adaptation to market changes and utilisation of small new possibilities on the market. Further, the chapter emphasises how this innovative role of the firm is implemented internally in the organisation.

This is a part of the general looking for a 'Schumpeter III' explanation that is the purpose of the book. Innovation as individual entrepreneurs' radical acts through firm establishment ('Schumpeter I', cf. Schumpeter 1934) has generally failed as the dynamic factor of the economy. Science-push and technological development as the innovative activities (which has been called 'Schumpeter II', cf. Phillips 1971) maybe also losing some of their dynamism. Thus, we should have a new understanding of innovation. Where could we find that?

This brings us to Kirzner's theory (1973). To a great extent innovation theory has focused on Schumpeter, particularly 'the young Schumpeter' (or Schumpeter I) (Schumpeter 1934). In a way he accepted the fundamental axiom of the economics of equilibrium, but stated that innovations from time to time completely destroyed the equilibrium and established a new one. Innovative behaviour (in his version in form of entrepreneurship) had the function of creative destruction. Kirzner rejected Schumpeter's idea of entrepreneurship as creative destruction which destroyed equilibrium. He also accepted the equilibrium axiom, but stated that there are several imbalances. To him the entrepreneur has the function of exploiting un-utilised market possibilities and in that way they adjust the economic system and make it function more efficiently. They restore the equilibrium. The entrepreneur is a responding agent that can see new possibilities and combine them, he is not the source of radical innovations.

Could Kirzner's theory explain innovative behaviour as it has been discussed here, i.e. emphasise a series of many small changes? An explanation of innovative behaviour could be developed from focusing on individuals – entrepreneurs – to focusing on changes and adjustments developed within organisations (a 'Kirzner II' approach). The extreme consequence of Kirzner's view would be that there is no economic development outside the adjustments of the imperfection within the equilibrium system and the idea of jumps through creative destruction is not valid. That would possibly destroy the basic idea of evolutionary economics and put the issue of innovation within the framework of the neo-classic equilibrium model emphasising the adjustment of imperfections. How could we then explain economic growth and the development of new fields such as IT hard- and software and services and the destruction of old industries such as producing horse carriages? Could the entrepreneur not create new markets?

A solution would be to combine Kirzner's and Schumpeter's views in an organisational setting. Firms make many adjustments to un-utilised market conditions, not as single entrepreneurs, but as a result of social organisational processes. However, these small adjustments or changes not only restore market imperfections, they also sometimes create new market possibilities that did not exist before. Mostly it is a mixture of imperfection restorings and small steps towards creating new markets, but sometimes (although rarely) these small steps meet and are combined with some radical inventions (such as the chip which led to the whole IT revolution).

This view will be the basis for the theoretical discussions in the chapter. First are the two core notions, strategy and dual organisation, presented and discussed. These will then be the basis for discussing some of the core issues in contemporary innovation theory: what innovation is, the influence of the environment on the innovation process, knowledge and entrepreneurship, organisational learning and technology. This is done to show how this view influences the total interpretation of innovation activities and thus fulfilling the overall intention of the book.

Strategy

As strategy is the core concept in this chapter it is necessary to define what I mean by strategy. The term has been used in many different ways (Mintzberg 1989, Mintzberg and Waters 1982, Chaffee 1985) thus it is not at all clear what it means.

Strategy will not be defined as generic market placement (as Porter 1980, 1985 does) according to which the firm selects a certain mode of development according to its market position and maintains that direction. In Porter's model this way could be either cost leadership, product differentiation or market segmentation. In their innovation activities, firms are market oriented as argued in this chapter, but they also have to take their internal forces – their resources and capabilities (cf. Teece and Pisano 1994) into consideration. Therefore, a pure market oriented interpretation of strategy as advanced by for example Porter is insufficient for this purpose. It is also a very rationalistic interpretation (cf. Knights and Morgan 1991). My studies in service firms (Sundbo 1998b) and some studies in manufacturing ones (cf. also Burgelman and Sayles 1986) show that the firm's strategies are not like that.

The definition of strategy here is that it is interpretative: the management makes an interpretation of the development of the market including the firm's market situation, its internal resources and capabilities. It is not possible to rationally deduce from an analysis of the market and the internal resource situation which strategy to select. Strategy is about the future. Since no objective truth can be found about the future (which here is a different assumption from that of the strategic planning approach), the strategy is a series of signs or indicators of several possible futures interpreted through the glasses of the management. One has to choose among several possibilities of an uncertain future. Strategising is thus not here thought of as a fully rational process, it is a quasi-rational process. This means that the management attempts to get as true a picture of the future market, the future competitors and the future internal resources as possible, but that it is not possible to get an objective truth. This interpretation process can not avoid being but a hermeneutic process (cf. Schutz 1967): the management lays its interpretation system as a filter over the strategising process. The interpretation may be developed by the top management alone or in an interaction between the top management and the employees and middle managers. The top management, or the managing director, however, has to decide upon which interpretation to choose, thus the top manager(s) are responsible for it.

As a starting point we may assume that the management's interpretation of the future possibilities is made as neutral and in the interest of the firm as possible. The managers attempt to be rational about the future, but it is not possible. This is also why strategies from time to time are changed when the future turns out to be different than thought.

Here, I am particularly interested in the innovation aspects of the strategy. This is a broader interest than just the question of whether the strategy includes goals for technology development (cf. Horwitch 1986) since innovation here is defined more broadly than just technological innovations. The formulation of strategy in firms takes its point of departure in the environment in the situation (described above) 'when the goals have to be established'. Innovation has the role of maintaining the existence of the enterprise in the future and, perhaps, grow.

The formulation of strategy is also a political process. Different actors within the firm – such as managers, shareholders, employees and others – can have conflicting interests and influence the strategy. This can lead to a complex strategising process. There is a possibility that the management can use the strategic interpretation as a means of power. Then the manager's or the managers' personal interest can be placed above the firm's interest. The strategy is at once an objective goal analysis and a potential power system. This implies that innovation is also potentially an object of power struggle.

The formulation of strategy is a process. It takes some time to develop it and one can never be sure that the chosen strategy is the best. Employees and managers are often involved in that phase. This view of strategy is thus processual (cf. Mintzberg 1973, 1989, Pettigrew 1985).

Even after the strategy has been formulated, it is a process. The firm must continuously relate itself to new behaviour by the competitors and other market actors, not least the customers. That is also why the strategy can not be a generic one. Employees and managers are also involved in that process because they often recognise new behaviour from the market actors first and because they may have valuable ideas for new strategic elements. Furthermore, new internal factors could lead the firm to change its strategy. That could be inventions, new technology coming from outside, new ideas from employees etc.

This means that the strategy is under continuous observation and will be changed from time to time. This is not to say that the strategy and development of the firm can be described as chaos (cf. Stacey 1993) and if it is, it is at least 'controlled chaos' (Quinn 1985). Firms may generally be assumed to have fixed goals for a period, however, they might change before the end of a scheduled period if developments of the market situation or internal resources and capabilities suggest this. Thus, the realised strategy at the end of the scheduled period may be different from the intended one (cf. Mintzberg and Waters 1982, Nyström 1979).

Of course, there are firms where the strategic situation is chaos, but it is assumed that they are exceptions. These situations could be of two kinds: (1) very innovative entrepreneurial firms in new areas (e.g. actual firms creating web sites); (2) firms with bad management (and they will not exist for long time).

It is important to establish the right goals at the right time, which should follow the market developments and the product life cycle. It is the task of management to follow these developments and a criterion of good management is that the manager knows when to change the strategy and thus the direction of innovation (which types of innovation should be developed).

The strategy becomes the framework for the firm's development and innovation. The goals set up in the strategy can function as an inspiration for innovation and specifies the direction of the firm's development so employees and managers can see which types of innovation fit into the strategy. However, the strategy also often functions as the management's tool for deciding which innovative ideas to accept and which not.

Dual organisation

Since innovation here is assumed to be determined within a strategic framework, which only gives general development guidelines, the process of getting ideas and developing innovations within the firm may be organised in different ways, according to the situation. However, the view of the firm's innovation behaviour as a complex process of small changes (and only rarely more radical innovations) in a strategic framework leads to a certain core model of the way of organising the innovation activities. It is called the dual organisation, which will be explained in this section.

This is the Kirznerian adjustment function in a collective setting (which means that more individuals are involved). The adjustments are made by a series of individuals in interaction processes. Ideas may often come from individuals, but they will soon be collectivised in the way that other people will react to them and the idea maker needs sponsors in the organisation (cf. Pinchot 1985). Often the idea will have been born from the interaction with other people inside or outside the organisation or the idea may be born by a group. Further, the development of the idea often demands that more individuals are involved. Generally, the development process will be carried out in project teams or other organisational settings, e.g. a particular venture department (Burgelman and Sayles 1986) or another existing department, even though they may have a very loosely coupled and flexible organisation.

Many small changes will not be organised in this way, but they will nevertheless normally be developed interactively. Even the more radical, Schumpeterian part of the general innovation process may be explained by such interactive processes between individuals in the organisation. The struggle for having the more radical ideas accepted is generally a political process where other members of the organisation must be convinced. Also, the development of the innovation, which includes solving a lot of problems on the way, demands interaction between several people.

The strategy is the guideline for the birth of ideas and further development into realised innovations and the top management is the guide who also make decisions, but the top managers can not get all the ideas themselves. Therefore, they are dependent on ideas coming broadly from throughout the organisation. Involvement of all or many employees and managers in the innovation process is necessary (empowerment, cf. Kanter 1983). The strategy can function as an inspiration for the employees to get ideas. However, all this can very easily be a confusing process and it therefore needs some regulation (Sundbo 1992).

This leads to the model of the dual organisation of innovation. It says that there is a loosely coupled interactive structure in which ideas and development of innovations take place in an interactive way (cf. Sundbo 1998b) and then there is a hierarchical managerial structure which regulates the innovation process. The interaction between the two structures is as follows:

- Ideas appear in the loosely coupled interactive structure.
- The management may create incentives for innovation thus inducing innovation (Binswanger 1978).

- The ideas are developed within the loosely coupled interactive structure.
- The management makes successive decisions whether to go on with the innovation.
- The people in the loosely coupled interactive structure may attempt to influence the management's decisions.
- Thus, an interaction process between the two organisational structures is created.
- The strategy is the decision framework because it is the guideline for the firm's development.

It has been empirically demonstrated that the dual innovation organisation is widespread in service firms (Sundbo 1998b, Sundbo and Gallouj 1999).

Sometimes corporate entrepreneurs from below may carry the innovation the whole way through to market realisation as a classic individual entrepreneurship process, but this may be assumed to be rare. So many problems will appear on the way and so much knowledge must be provided that it is currently very difficult for one person to carry through such a process. When a classical entrepreneurial type with a radical innovative idea appears, he or she will often break out and establish his or her own enterprise.

Mostly several people are engaged in the innovation development from below. They are normally not entrepreneurial personalities, they play roles in the interactive innovation process (cf. Sundbo 1998b). One crucial role is that of a corporate entrepreneur (or intrapreneur cf. Pinchot 1985), which is entrepreneurship carried out for a limited time in a certain situation. Also other roles, which may be temporary, exist: for example, the role of the analyst, who has an important function in the process of strategy formulation, and several roles in the project teams that develop and implement the innovations.

Innovations can also be top planned, long-term activities which are carried out in institutionalised organisational settings, normally an R&D department, but it might be in other ways such as establishing an independent venture department or firm (cf. Burgelman and Sayles 1986). This will generally be after an idea phase where several people are involved outside this specialised department. The top management will have been involved in the decision process for which the strategy – the reflections of the market situation and internal resources and capabilities – will be a core decision framework.

However, the assumption here is that such a linear top-down process through a specialised department (such as an R&D one) will be rare, even in high-tech areas. We have an example from our research about how a Danish pharmaceutical firm thinks concerning innovation. The research manager explains that the firm takes its point of departure in the market and the future wants and problems that the customers may have. They create a strategy which not only focuses on the clinical disease diagnosis, but also on the daily behaviour of the potential clients. The circumstances (e.g. side effects, the daily life routines etc.) are also important, besides solving the core problem (i.e. a clinically diagnosed disease). Then they set up a complex structure of project teams, informal interaction processes etc. to find information and get ideas. The R&D department is centrally involved in the process because a lot of technical research and development is involved, but the process is not completely left to that department. In some cases the specialised R&D department may break the strategy by getting an idea outside the strategic lines and policy and perhaps start the development process before it submits anything to the top management. An R&D department is also a political actor within the firm. This situation may be supposed to be even rarer. Further, the innovation may then be rejected by top management or it will lead to a change of strategy because top management can see that the innovation can create a new market or market segment (the latter, for example, by a quality improvement or productivity increase).

Innovation and change

In the previous sections contemporary innovations have been stated as adaptation through small changes. Firms' development, and hence the dynamism in the economy, is characterised by many small changes in many dimensions. Examples are: a new bank product – a credit card which can be obtained within 15 minutes. That idea has come from market analyses. It also demands risk analyses and even a new risk assessment system. New technological possibilities have also been a determinant. The concept came from a group of employees after a process of involving all employees in idea-generating activities. An engineering consultancy company operates through project groups. Innovations are made as adjustments in each project team. The adjustments can be determined by different factors: special customers wants, new technological possibilities, the organisation does not fit into a new task etc. Almost each project team makes some small changes that, combined, develop the firm so it can keep its competitive position. Firms producing hearing aids can no longer compete only by technological improvements. They must also take into consideration the users' aspect (practical life of the users), market image etc. They transform their organisation into a highly flexible one where each employee is supposed to get new ideas of many kinds. Innovations crucial to the firm development become of many kinds: innovation in products, processes, organisation, market behaviour and delivery and new raw material (cf. Schumpeter 1934); technological and social innovations etc. Further, the Kirznerian approach emphasises the market adaptations.

This challenges the innovation concept. In recent literature, innovation has generally been understood as a distinct change that can be clearly identified and delimited and which causes a jump in turnover or profit which can also be clearly identified. In a situation with a series of different, but together-woven, small changes it is not easy to identify and delimit single changes. Nor is it easy to point out the effect of the single change to the turnover and profit, but together all the small changes increase turnover and profit.

The environment and the innovation process

Markets become increasingly turbulent and complicated to operate in, particularly as they become global. Many markets are satisfied, which makes it more difficult for a firm to grow and even maintain the existing business. Further, the competition increases in most markets. These developments demand more dynamic and innovative behaviour from firms in order to compete because the competitors will innovate. However, it is not necessarily wise to maximise the innovation process. That could cost many resources, which could be wasted. In many cases it is clever policy to wait with innovations until the market for one's own products become satisfied or the competitors start innovating (cf. Kotler 1983). The issue of when to innovate and who starts to do it is a game. Strategy is the firm's conscious setting of goals in that game.

Two conditions in the environment are decisive:

1   *When the goals have to be established*: This is a question of expectations about future developments. What will the needs of the customers be in the future? What will they demand (which does not need to be synonymous with their needs)? How will other actors (competitors, regulatory state authorities, consumers' movements etc.) react and develop their policies? How will the employees develop concerning interests, competencies and motivation? Which new technological and organisational inventions (or trajectories) will appear? Which new internal resources and capabilities does this demand?
2   *When the goals have been established and innovations must be developed*: This situation raises questions such as: How do we procure knowledge? Do we have the right resources and capabilities – and if not, how do we procure them?

The first condition is under-prioritised in innovation theory. There has recently been a great deal of attention given to the second condition. The knowledge and learning economy (OECD 1996) has emphasised external knowledge procurement. The resource based view of the firm (Penrose 1959, Grant 1991) has been applied to innovation (Teece and Pisano 1994), which has placed greater emphasis on internal resources and capabilities. The first condition has, of course, been emphasised within marketing literature (cf. Kotler 1983), but the innovation issue has not been so strong within that literature, nor has the tradition developed comprehensive models of how the innovation process should be organised. My case studies in innovation in services show that the first condition is the most difficult to handle. The firms often know how to procure knowledge, resources and capabilities or they learn how to (although it may not be perfect and they can learn to do it better). Knowledge about which direction to go in the future and when to innovate is much more difficult to procure and they have less experience in that. Nevertheless, they are forced to do it. Inactivity is also a strategic choice, even if it might be unconscious.

The environment is not only important in the phase of strategy formulation. It is also important in the innovation process. External factors are included as the innovation literature demonstrates (von Hippel 1988). The employees and managers interact also with the external world.

The external factors are of two kinds (cf. Sundbo and Gallouj 1999, 2000):

1   *Trajectories*: These are principles and logics of certain development lines. Dosi (1982) has introduced technological trajectories as certain basic logics of technological development which lead to many derived innovations, e.g. the steam engines as basis for mechanisation, the ICT revolution based on chips and algorithms etc. This can be extended to other trajectories, e.g. social (social movements in society which influence the customers' attitudes), institutional (such as creation of the European community and the single market), managerial (basic management ideas such as taylorism, motivation, business process reengineering) or service professional (methods in service work which is not technological, e.g. insurance knowledge such as risk management). The trajectories are the knowledge highways to the environment. They identify the knowledge procured for carrying out the innovation process.

2   *Actors*: These may be customers, competitors, suppliers and the public sector (which may regulate the field in which the innovation is placed within). Also the shareholders of the enterprise are important actors in the innovation process. The actors are important because they have knowledge, but also because they are powerful political actors. Their attitude and behaviour can be decisive for the success of the innovation. The firm will therefore be forced to take that into consideration and attempt to make them positive to the innovation and eventually make alliances with them or involve them as collaborators in the innovation work. In that way innovation is a game with competitors and potential allies or enemies.

The strategy will still be the guideline for the management and employees of the firm in their attempt to find trajectories that can give knowledge input to the innovation process and to make alliances and co-operations with external actors. It is like in war games – from whence the concept strategy comes – a game between parties although in innovation there are more parties.

The environment is also involved in the innovation process because the crucial factor is whether the innovation will be accepted by the market. Even if it is a process innovation will often have some impact on the firm's market position, for example decreasing the price of the products or even if it has not, the firm is dependent on not having negative comments in the press and negative reactions from consumers. Continuous reflections are made and innovations are normally tested on a consumer panel (at least if it is a product or market innovation). This also means that the marketing department often gets a core position in the innovation development process.

Knowledge and entrepreneurship: innovation is an interactive process

How is the knowledge produced and used within the firm? Externally produced knowledge is mixed with existing internal knowledge. Recent innovation theory has focused much on knowledge as a core factor in innovation (OECD 1996). In extreme cases knowledge is considered as a determinant. That means that new knowledge in itself can lead to innovations (Mensch 1975). Learning has been seen as a prolongation of the knowledge view. Economists have spoken of the knowledge economy (Machlup 1980) and even the learning economy (OECD 1996), learning about innovation has been seen as the ability to more efficiently select knowledge.

Knowledge can in that framework easily be understood as being rational. It is within the tradition of economics to find rationality in all activities, also for measuring indicators, which certain economic organisations such as OECD and Eurostat attempt to do by measuring investment in universities and other knowledge producing institutions, R&D investments, innovation survey indicators etc. However, it is a question whether this approach is in accordance with real innovation processes. The entrepreneurial aspects (e.g. Sexton and Kasarda 1992) have been excluded. These aspects emphasise phenomena such as idea generation and creativity (which are not equal to procuring knowledge – sometimes on the contrary), the will to fight for realising one's ideas, a drive to get a social position through innovating etc. They are not rational aspects in the traditional economic way of understanding the matter. Nor are all creative elements. Many of them are psychological and interactional, i.e. they are developed in interaction between people.

One example of this is how banks and insurance companies get new ideas for innovations, cf. case studies that I have done (cf. Sundbo 1998b). Employees were the most common source of innovative ideas, then came the customers as suggested by modern innovation theory and service management theory in particular. The important thing is how the ideas come from the customers. It is not because of any existing rational knowledge of the customers' needs in the future. The customers did not even have clear ideas of what they wanted for the future. The idea generation was, in most cases, the result of interaction processes. The customers had certain problems or vague wants which were observed by the employees in their daily collaboration with the customers. On that basis the employees developed ideas for new products or at least a clear definition of the problem that could be the basis for future innovations.

Entrepreneurs often have an intuition for market possibilities that can not be translated into rational knowledge. They can choose a successful direction that could not be found by rational analyses. One may attempt to dissect their psychology to see if one can find the elements, but often it seems to be the drive to fight and follow their own lines of development, no matter how the environment reacts and no matter how many defeats they face.

These entrepreneurial and creative elements must be emphasised in innovation theory. It is not that knowledge is unimportant to innovation. It is extremely important, but not isolated as a 'stock' that can be taken out when necessary. It must be seen in relation to creative and entrepreneurial elements. Knowledge is not a determinant of innovation. Several studies have demonstrated that entrepreneurship is not produced by knowledge institutions such as universities (Sexton and Kasarda 1992); generally, on the contrary. The driving force of innovation is something else – the will to innovate, creative, and entrepreneurial elements. Knowledge is a core resource that can be activated to develop the innovation when one has decided to innovate. It may also be the reason to decide to innovate, but it is not a determinant in itself.

The argument for knowledge determined innovation can not be saved by referring to the separation of codified and tacit knowledge (Polanyi 1958). What I am talking about here is not tacit knowledge. That is a part of creative and entrepreneurial behaviour, but only one element, and it is not the one that determines the decision to be an entrepreneur. The phenomenon of getting an idea, being creative, and an entrepreneurial fighter is not only determined by one's tacit knowledge.

Creativity initiation, entrepreneurial fight and knowledge development happen in the dual organisation. Innovation processes can be bottom-up, i.e. starting in the loosely coupled interaction structure, or they can be top-down, i.e. initiated by the management.

Individual creativity and entrepreneur characteristics are important determinants of innovations, particularly when they come from bottom-up in the organisation. Knowledge is a core factor and could be a determinant if the innovation comes top-down (either decided by the top manager or developed by the R&D department), but that is assumed to be rarer than the bottom-up process.

Organisational learning

One core issue in economics and business administration is how processes could be made more efficient. That means with lower costs and better results. Even if the question can not be answered in an exact and deterministic way (particularly not concerning innovation), it is a behavioural parameter behind the management's organisational initiatives. Since innovation is largely composed of a complex pattern of many small changes and, sometimes, a few more radical innovations developed in an interaction system, it is very difficult for the firm's management to ensure the efficiency of this process. It is not sufficient to have a standard model read out of some book. There are many successes and more failures in the many change processes in the firm. The experiences are spread and the top managers do not know all of them. Further, knowledge and competencies are spread among employees and managers.

This leads to the idea that organisational learning is the way to ensure improvement of the firm's innovative behaviour (Argyris and Schön 1978, Senge 1990), particularly within a Kirznerian adaptation setting. It is important that the knowledge of the experiences of the many innovative acts and interactions is collected in the organisation and systematised so the rest of the organisation could learn from them. Learning is generally individual and much knowledge in firms is tacit. This makes it difficult to collect the experience as organisation learning, i.e. that the experience is present in the organisation even if the individuals disappear. To the degree that organisational learning is possible, it takes place within the dual organisation. That demands a conscious politics from the management with the following elements, which should induce learning contributions from the individuals through the loosely coupled interaction structure:

- Inducement of individual, team-wise and departmental learning
- Codification of the individual experiences and knowledge
- Communication of the individual experiences and knowledge
- Storing of the individual experiences and knowledge
- Creation of a distribution system so all members of the organisation have access to the knowledge stores
- Incentives for the members of the organisation to use the knowledge store
- Storing of experiences about how to best organise this organisational learning process

Organisational learning, particularly about innovation, has shown to be extremely difficult for firms to achieve, the longer down the above list one goes (Sundbo 1999). Learning is often strategy-specific, i.e. it is specific for the circumstances of one strategy. If the firm changes the strategy or one attempts to transmit the experiences to another firm in another situation, many of the experiences are not valid.

Technology

Technology has been a little suppressed in this discussion due to the fact that I have wanted to emphasise the strategic aspects in general and technology is only one element in the innovation processes. However, technology is an important element – even within a strategic and interactive understanding of innovation. Many products are technological – all in manufacturing and many in services as well. This will not be discussed here. What I will discuss is technology as a strategic factor: how technology is involved when the firm formulates a strategy.

Technology can strategically play a role in the innovation process in the following three ways:

Trajectorial

The firm follows a certain technological development, a paradigm or trajectory (cf. Dosi 1982, Perez 1983) which has a certain logic and certain development of knowledge. This logic and developing knowledge leads to a series of new technological elements, which can be the basis for innovations. One firm may relate to several technological trajectories.

This role is particularly distinct within scientific based professional fields such as pharmaceutical or ICT (whether hardware or software). There are many knowledge developments outside the firm that the firm can rely on and internalise. It provides a possibility for a more or less permanent innovation process, but it also creates a pattern dependency which can be difficult for the firm to break because the professional trajectories are so strong.

In this situation the technology trajectory is superior to the strategy, or – in other words – the strategy begins to follow the technology trajectory. Likewise, the organisation of the innovation activities becomes establishment of professional teams or departments, often an R&D department.

Eclectic

In this situation the technology elements are selected from many sources and trajectories. There is very little technology push as is the case in the first situation. It is a pull situation where the strategy is the superior, guiding factor. The management decides the goals for the firm's development and the strategy. Innovation ideas and concrete projects are formulated and the project teams, or other responsible units, collect the knowledge necessary to solve the problems that appear throughout the innovation process. This may be knowledge about human behaviour (e.g. customers' or employees' daily behaviour) or others, but it may also be about technology. However, it may concern very different technologies such as chemical, ICT, transport, cleaning and clothing technology if we think of a complex chemical press (e.g. production of pharmaceutical) or just a laundry (which for example washes clothes from the pharmaceutical industry).

The selection of technology is very decentralised, often handed over to project teams or single persons. Technology is not so much a determinant as in the first situation. The danger of the eclectical form is that the firm misses relevant technological knowledge because it does not know of its existence and does not find it.

Fundamental

One basic technology becomes fundamental for a firm such as IT hardware for a software or hardware firm or milk manufacturing for diaries. This basic technology provides many possibilities for the development of concrete innovations. In comparison to the first situation the firm here has a freer selection and development of technology and it is not pattern dependent. The firm collects new knowledge from the environment and thus enters the technological trajectories, but this is only one type of input. The firm may internally develop new technology and innovations just as much or more so. Since there are many possibilities within the framework of the basic technology, there is no determination of the direction of development. The situation demands a strategy concerning which direction to chose.

This situation is as open as the eclectic one, but the basic technology is more decisive for the strategy. The goals must be based on developments that the basic technology makes possible. This situation can create survival problems if the basic technology becomes mature and the firm can not move itself into one of the other two situations or introduce new productions based on other basic technologies.

In all three situations technology is not the only determining strategic factor. Other factors such as market possibilities, knowledge about peoples' behaviour, organisational factors etc. also play a central role.

Case examples of strategic innovation situations

To illustrate the model of strategic innovation I will briefly present two cases and discuss them in relation to the model as developed in the former sections.

Small bank

The first case is a small Danish bank with about 275 employees (the case is described in Sundbo 1999). It is mainly a retail bank, having only a few business customers (of which the main part are pension funds and associations). The bank has a very dynamic managing director and has grown within the last fifteen years. It has, in particular, competed on the price policy: low loan interest rate and high deposit interest rate.

Strategy

The bank has formulated a market segmentation strategy: it wants the better off private customers. That is the reason why they can have a low interest differential since they have very few losses on that customer segment. Recently, they have made a strategy to increase their number of business customers. They have realised that they must be in front concerning new services and other types of innovation if they wish to exist in the future because the large banks have decreased their interest differential so they can not continue competing on that factor. They also want to be in front regarding the use of IT in form of Internet banking (or home banking).

A part of the strategy is to develop the employees and the organisation towards a self-generating learning organisation, i.e. where the employees and managers systematise the experiences and create innovate ideas themselves.

The internal resources and capabilities – which are the employees' competencies and the IT capability – are a core part of the strategy.

Innovation initiatives

The bank has connected innovation to a general organisation development towards a flexible and learning organisation. The goal is to have more motivated employees who involve themselves in the bank. They should also be more flexible concerning working tasks, thus they could help each other. The employees and managers have been involved in reformulating the bank's strategy and the goal is that they also should be active in innovation processes, already as idea makers.

Four training programmes were established. One trained the employees and managers in strategic analysis and the outcome should be a new strategy. Another was a training programme for the managers about learning organisations. A third was a programme of personal development for the employees. The fourth one was a group oriented innovation programme: It was announced that all employees and managers could present ideas for innovation and they could then establish a team to develop the idea; the team could use a part of their working time for this purpose.

Organisation of the innovation activities

A real dual organisation was established. Through the training programmes all employees and managers were requested to act as corporate entrepreneurs: to present innovative ideas and develop them and to relate this to the strategy, in which formulation they have been involved themselves. Innovative ideas were also the outcome of the strategy training programme. Several innovation teams were established. However, the managing director wants to decide everything concerning all ideas and they are presented to him at different stages of the innovation process. He has rejected some ideas and accepted some according to the strategy – and his personal preferences. Even though these preferences naturally play a role in the decision process of the top manager, he needs to keep the ideas – even his own – within the frames of the strategy, and he does so. If not, it will destroy the trust in the whole development project and the involvement and motivation of the employees will disappear.

Thus, much corporate entrepreneurship has developed and employees and managers have been really enthusiastic in participating in the innovation processes. The innovations have been based on very practical ideas that have not demanded very much or very complex formal knowledge, which must be provided from outside – as often is the case in service firms (cf. Sundbo 1998b). When it comes to the next step, development of the innovations, it has demanded some formal knowledge, particularly the IT-based innovations, but it has also demanded explicitation of the internal tacit knowledge (cf. Nonaka and Takeuchi 1995). Technology as IT is central to many of the innovations, but several ones are not technological (such as a new way of treating customers in person-to-person contact).

One can not say that external factors are much involved in the innovation processes in the bank. It is member of a data processing central (jointly owned with other banks) and this one has been a core actor in the development of the technological parts of the innovations. Factors in the environment have been a basis for the innovations, but primarily a basis for ideas that the single employees get, not as actors in the development of the innovations.

The bank has gone far in developing a learning organisation. However, mostly in creating an innovative culture and concrete innovations (double and single loop learning according to Argyris and Schön 1978), not that much in collecting and codifying experiences of how to organise the learning process (learning to learn – deutero learning according to Argyris and Schön).

Concrete innovations

It is possible to identify the concrete innovations since they have been organised as projects or ideas for projects. They have been of different kinds and sizes (whether innovations or just small changes), but all realised ones have been widespread in the organisation. Examples are: a new customer analysis system that can be used for marketing; a new culture for interacting with customers in person-to-person contact; a new credit card which can be delivered within 15 minutes (where the innovative part is in the risk assessment system).

Engineering consultant

The second case is a Danish engineering consultantancy company. It has about 2200 employees and is among the larger ones in the world. It simply sells advice and does not act as the building enterprise contractor. It is a marketing consultancy in several technical areas within three main fields: (1) building operations, (2) environmental problems, (3) traffic systems. Lately, consultancy within economic and sociological fields such as public administration, organisational development, economic management etc. have become a part of the product portfolio. The firm exports more than half of its production, primarily to Eastern Europe and developing countries.

Strategy

The firm has had a fairly conservative strategy since the business within the existing fields has gone well and the market has grown. It has only had to maintain and strengthen the core competencies. However, lately the market has changed with more players and increasing price competition. The consequence has been falling profit. Therefore, the situation calls for a new, innovative strategy. The firm should be more offensive in expansion abroad, which seems mostly to happen when the home market is under pressure. Process and organisational innovations should be developed to minimise costs and product innovations could advantageously be introduced. These demands are admitted by many of the managers, but they have not yet expressed them in a formal strategy.

Innovation initiatives

The firm organises R&D activities in form of cross-departmental teams that shall develop innovations. It concerns technological issues such as concrete technology, energy technology, waste water technology etc. Several patents have been taken out. The innovations and patents are not technology in the meaning of concrete objects or goods, but the knowledge of how to make such objects, for example computer simulation models and methods. This has been a long tradition in the company. Many changes are results of the consultancy services. They are not standardised, or they are semi-standardised. This means that each customer and each case has its own problems which demand new solutions. This is considered as a part of normal consultancy work.

Organisation of the innovation activities

This follows a pattern that is traditional within the firm. As mentioned, it is composed of two parts. One is the R&D projects which has the purpose only of developing innovations, not concrete consultancy. These projects

are decided by the top management and they are organised as cross-departmental project teams. Different types of professionals can participate in this
work, including for example economists and sociologists, although normally engineers are the participants. They bring with them the tradition
of science based R&D activity known from manufacturing.

The consultancy work is normally organised as project teams. In the
work with solving the problems of the customers they develop new solutions. The customers are often others than the users. Since the engineering
company works much in Eastern Europe and developing countries, those
who pay for the projects will often be foreign governments or international organisations such as the UN, the EU etc. while the users are local
authorities or populations. This creates extra problems that call for unconventional solutions. The company also seeks unique tasks to develop this
competence of developing unusual solutions to new problems. This
demands a bottom-up entrepreneurship and the company thus requires
employees who are development and innovation oriented.

Knowledge and knowledge based competencies are core factors since
that is what the company sells. Further, its fields are within scientific and
professional traditions where abstract knowledge has always been crucial.
The R&D projects are generally science and knowledge based while the
concrete solutions in projects may be more characterised by entrepreneurship. Creativity and quick ideas based on practice experience is
important, besides abstract knowledge. Thus, concerning the relationship
between abstract knowledge and practice based entrepreneurship the engineering consultancy company differs from the bank.

Technology of course is a core factor. Not all the products are fully
technological. Some are social (such as economic management systems,
municipal administration systems etc.), but most products comprise advice
concerning technology. IT is process technology which is essential for
development of the projects and for the administration. External factors
are central for the innovation process, not only as idea bases. The company
attempts to involve customers as partners in development projects and
thus in the innovation processes.

The engineering consultancy firm is not good at organisational learning.
The experiences become tacit knowledge in the heads of the employees
because the work is non-standardised. The project form implies that the
employees disappear to a new project before the former is completed and
there is no time for summarising experiences and giving them to other
people. The management sees this as a problem, but has not yet developed a plan for what to do about it.

Concrete innovations

Examples are given above. It mainly concerns models and methods of the
technical activities that the firm is giving advice about. A few social product

innovations such as administration systems etc. come out, but this a new field so there are not very many. Some process and organisational innovations may also come up, but they are rare. Many of the innovations are small changes that happen in each consulting project and which may be unrepeated. It is therefore to a large degree difficult to identify the innovations and many of the activities should be characterised as a continuous, complex change process.

These two cases are equal in many points: innovations are strategically determined, influenced by the firms market situation as well as the internal resources and capabilities. Many different types of innovation are developed. The organisation of the innovation activities is dual: a combination of bottom-up entrepreneurship and top-down guiding of the innovation process. These are the main points of the model of strategic innovation. The cases also demonstrate that some aspects vary from firm to firm: the use of technology, abstract knowledge in relation vs. entrepreneurship and how much the organisation is a learning one.

Conclusions

I have discussed an interpretation of innovation as a strategic process. A conclusion, which outlines a model of the strategic innovation process, may be summarised in the following points:

*   The environment, particularly the market possibilities, is crucial for the firms' innovation behaviour.
*   The environment can neither be exactly explained nor controlled.
*   Therefore, the management of the firm makes an interpretation of the external situation and where the environment is moving towards.
*   This is combined with an analysis of the internal resources and capabilities and some general goals for the firm's future development are set up in the strategy.
*   The strategy guides a broad innovation or change process within the firm.
*   In this change or innovation process is the employees involved as idea makers and corporate entrepreneurs.
*   However, the process is guided by the management, thus a dual organisation appears (a dialectic between a hierarchical top-down process and a complex, somewhat anarchic, bottom-up process).
*   The strategy is only fixed for a period. It is also a process where new knowledge or the manifestation of different actors may change the existing strategy.

The model describes the situation in most service firms very well and it is assumed that it is also adequate to describe the situation in most manufacturing firms. However, to settle the latter demands new case studies where the innovation process is seen from this perspective. Existing studies can, in most cases, not be used to investigate the issue because they are done from another theoretical perspective.

The contribution to a new understanding of innovation

In relation to the overall aim of the book, this chapter has discussed some theoretical elements.

First, it has contributed to a specific interpretation of innovative behaviour. Innovation is seen as a continuous process of small changes. The firms' behaviour may be described by Kirzner's function of coordination and utilisation of market imperfections (Kirzner 1973) more than Schumpeter's (1934) function of creative destruction.

The chapter has further contributed to an understanding of innovation as reflexive strategy by discussing of strategy as the 'riving shaft' of the innovation process. The firm's innovation behaviour is decided by movements, problems and possibilities in the environment. These are interpreted by the managers and expressed in the strategy, which becomes the guideline for the innovation activities. The strategy has a double function: it expresses the possibilities and the way to go in the environment and it is an internal management instrument to guide the innovation process in the organisation. It is necessary to have such a guiding instrument since this process is a complex interaction process. The innovation organisation is a dual one: innovations may come from interactions between employees or from the top management and the interplay between the two structures is important to the process.

This theoretical view also has implications to the understanding of the role of environment, organisational learning and use of technology in firms, which have been discussed from a perspective of strategic reflection.

Part II

Externally oriented innovation activities in the firm

The understanding of innovation within the framework of strategic reflexivity, as it has been introduced in this book, begins by examining the environment in which the firm is situated The firm relates itself to this environment and the directions in which it might develop in the furture. 'Environment' is understood to mean, primarily, the market, but it also includes other societal aspects (e.g. political regulation, citizens' political and ethical attitudes, new management philosophies and new technological trajectories) which are important because they identify the future market. By 'strategy' we understand the creation of a direction for the firm's development given the anticipated environmental change. The firm's development will be limited by its innovations. Strategic reflexivity originates here, and the firm's relation to its market – and more broadly the environment – becomes the starting point of the detailed analysis of firms' innovation activities within this framework. This section thus introduces a more detailed, specific and profound analysis, which is more empirically based. What are the conditions for firms to formulate an innovation-oriented strategy? What is the strategic interpretation of the environment? Is it always a rational, smooth management process – or can it be a complex, contradictory one? These are some of the questions to be discussed in this section.

Harry Nyström and Sten Liljedahl argue that the most efficient innovation behaviour is an open, reflexive strategy, which means that firms do not follow a planned route of innovation. They also argue that both technological and market innovations are crucial and firms can not rely on just one of them. It is important that the firms are open to market moves, combine different technological elements (not simply following one trajectory) and be flexible in their innovation behaviour, this is particularly so concerning the process of low-tech firms transforming themselves into high-tech ones. They have studied thirteen enterprises, which have moved from low to high tech. The authors demonstrate that the most successful of them followed these rules. The firms launched different strategies concerning the market and technological innovations mix. The authors have created some general strategic models based on this mix.

Andrea Gallina analyses the situation and role of small firms. He discusses theoretically how small firms are woven into networks, which increasingly becomes international. The small firms become dependent on knowledge to innovate. Large firms, which participate in the networks, possess that knowledge. The small firms have many internal and external sources for procuring knowledge, but they have a particular symbiotic relation to the large firms, which have knowledge about materials, design, market etc. while the small firms have knowledge about the production process. The small firms' innovation processes are characterised by strategic reflexivity, but also by a power relationship where the large companies have the power in the networks. Gallina analyses small firms' situation in the Danish economy, which is an internationally open one. Case studies have demonstrated that the small firms mostly implement small changes.

Paul Trott emphasises the firm's dilemma between basing their innovation activities on technology push and market research. The first may be risky because it demands high investments and it is difficult to create discontinuous technology development. The second is in accordance with the idea of strategic reflexivity; however, Trott discusses the dangers of relying too much on market pull through market research. He puts forward several arguments to support his critical view: The customers do not understand new technological products thus market research prior to innovation is of low value. Customers stick to the well-known technology even if a new one is functionally better. Trott argues that often market research hinders radical innovations because it shows that customers do not want the new product. It limits creativity and erodes the entrepreneurial, fighting spirit. The chapter thus places its focus on the innovator–customer relationship.

5 From low-tech to high-tech

Technological and marketing strategies for finding new markets and technologies

Harry Nyström and Sten Liljedahl

Introduction

In this chapter, innovation is seen as the successful conjunction of technological opportunities in the environment, changes in the market place and in society. The basis for this is reflexivity, not planning, as is usually assumed in traditional approaches to strategy formation in the management literature. This is an evolutionary Schumpeterian approach to innovation management, which focuses on the creative way in which technological and market forces interact over time in creative destruction and construction.

Strategic innovation management then becomes a question of reflecting on and creating favourable indirect conditions which enable change, rather than following direct guidelines or normative rules, as in a planning perspective. Both in relation to other actors and consumers, and to the growing body of inter- and intraorganisational knowledge, firms must use open technological and marketing strategies. This is in order to make possible, and to facilitate, technological and market innovation in an uncertain world. To move from low-tech to high-tech, companies must see opportunities in a flexible interaction with the environment and favour a flexible rather than a fixed view of technological and market possibilities.

In this type of creative management framework, external orientation – (flexible cooperation with other actors), and synergistic technology use – (combining technologies in new and previously unanticipated ways) are seen as the main dimensions and driving forces of open technological strategies. By creating a potential for constructive change in products and services they are also a prerequisite for successful market innovation, by making possible open marketing strategies directed towards the fulfilment of new individual and societal needs.

Theoretical background

In the product development literature (Urban and Hauser, 1980, Wind, 1982, Cooper, 1986, Crawford, 1991, Kuczmarski, 1992) most models are mainly concerned with product innovation, how to compete with established product lines and technologies in given industries. Industrial innovation, entering radically new markets and industries by using R&D to transform, rather than modify products, is dealt with to a much lesser degree. In recent years, however, increasing attention is given in the literature to more high-tech product development situations (Gupta *et al.*, 1985, Green, 1991, Ng *et al.*, 1992, Wheelwright *et al.*, 1999) based on managing and exploiting technological change to find new markets.

The dynamics of moving from low-tech to high-tech product development has not been dealt with, however, to any large extent, despite the fact that this is the most pressing strategic issue for many companies, threatened by radical environmental change. Particularly in the case of commodity based companies, for instance basic chemical companies and agroindustry firms, we find many examples of companies faced with this need, due to overproduction and stagnation in the demand for traditional products.

A strategic development model

In the following, we will discuss, based on an empirical study of thirteen Swedish biotechnical and food processing firms, how companies may achieve this type of strategic transformation from low- to high-tech, that is from lower to higher levels of technological and market innovation in their new products.

For this purpose we will utilise a strategic development model, based on research carried out since 1975 in a wide range of industries and technology areas in Sweden (Nyström, 1979, 1989). The results of these studies indicate that more open strategies tend to lead to more unique and competitive products, both from a technological and market point of view. These strategies emphasise networks (external research and marketing cooperation) and synergistic technology use (new technological combinations)

Our model of product development is based on two main outcome dimensions (Figure 5.1), the level of technological innovation and the level of market innovation for new products. In both instances, these measures are derived from interviews with company representatives, who have been directly involved in their development

The level of technological innovation – our measure of technological success – is defined as the degree of creativity companies have had to employ to solve the critical technical problems when developing new products. The underlying assumption is that achieving a relatively high level of technological innovation is usually a necessary, but not sufficient, condition for achieving a high level of market innovation – that is highly unique and competitive new products – particularly in research and technology intensive industries. The more radical the need for creative thinking, in a specific development project, the higher the level of technological innovation.

We have chosen the level of technological innovation as our main indicator of technological success, since it is a more direct and inclusive measure

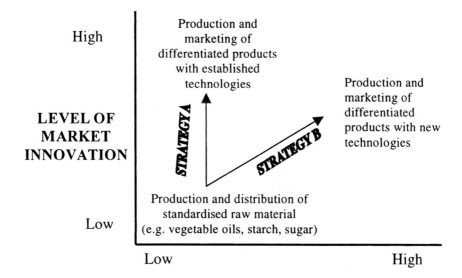

Figure 5.1 The product development strategies

than other more operational measures often used in the literature, such as patents or development time (Basberg, 1987). Patent protection is an imperfect indicator of the level of technological success, since for competitive reasons companies often do not apply for patents and therefore should not be used as a single indicator. Development time is also an imperfect indicator, since it reflects effort applied, rather than success achieved. The relatively high correlations in our data of these more indirect measures with our more valid, but less operational measure, based on the direct questioning of company researchers, give us greater confidence, however, in our direct main measure of technological success.

The level of market innovation – our measure of market success – is defined as the degree of market uniqueness for new products. The less interchangable a product is perceived to be by buyers when it is introduced onto the market, the higher the degree of market uniqueness. The assumption is that the more unique it is – the more that buyers see it as different in features and performance from competing products – the greater the market potential. An implicit assumption, which finds support in our previous research, is that companies do not usually introduce highly unique new products if they have not (by pre-testing or market research) convinced themselves that customers value the ways in which a new product differs from existing ones.

Examples of products with a low level of both technological and market innovation, as we have defined these terms, are, for instance, standardised raw materials, such as sugar and grain. With regard to this type of product, efficient production and distribution is the management's main concern. Prices tend to be highly competitive and overproduction often leads to surpluses, as we see in the agricultural sector. Supply and demand on the open market are usually too erratic to predict. This often leads to attempts by companies to control sales, for instance by using defensive strategies to tie existing customers more closely to their products.

To gain greater sales and profits, product innovation, i.e. market and technological upgrading of products, is usually viewed as desirable by low-tech companies, dependent on products with low degrees of technological and market innovation. To achieve this objective and become more innovative such companies need, however, to develop their technological and marketing strategies. This, then, also requires that they increase their innovative potential by designing and implementing a more creative culture and climate and more flexible forms of organisation, production and distribution.

This presents formidable obstacles to change for most low-tech companies. One of the most common ways for these companies to achieve a competitive advantage – which is shown in Figure 5.1 by Strategy A – is to try to increase the level of market innovation, without having to significantly increase the level of technological innovation. Essentially this type of low-tech marketing strategy requires companies to become more market oriented, by differentiating and positioning their products to achieve competitive advantage (Porter, 1985). Thus the companies, if successful in their efforts, may increase their degree of market innovation, by finding market niches, where their products are viewed as superior to competing ones.

Research intensity is, in general, very low in this type of company and mainly limited to perfecting existing technologies, rather than developing new ones. Technological strategies are usually very closed and mainly based on in-house development and refining established knowledge, which in our framework is called internal orientation and isolated technology use. Marketing strategies are also relatively closed and directed towards existing customers and modifications of existing products. This type of relatively closed development strategy is, for instance, almost the only strategy used by Swedish food processing companies (Nyström and Edvardsson, 1982) and the main development strategy for companies in the Swedish forest industry (Nyström, 1985).

A more radical and innovative development strategy, shown in Figure 5.1 as Strategy B, is necessary, in our framework, if companies are to achieve a high level of both technological and market innovation for new products. This strategy involves both high risk and strong possibilities for growth and profits. As our earlier research shows (Nyström, 1979), this usually requires that companies have relatively open and flexible technological and marketing strategies and high research intensity to generate new knowledge.

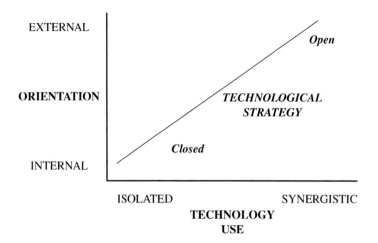

Figure 5.2 The main dimensions of more closed and more open technological strategies

More open technological strategies (Figure 5.2) are characterised by a greater degree of external orientation (more research cooperation) and more synergistic technology use (creative combining and recombining of existing knowledge). Strategic ventures with buyers, universities, other firms or inventors are often used to achieve synergistic technology use and external orientation in development activities. Some companies in this high-tech category devote upwards of 20 per cent of sales to R&D, while low-tech companies often spend less than 1 per cent. More open marketing strategies (Figure 5.3) are those which are directed towards new basic needs and/or new customer groups. The most open strategies are those which fulfil both these criteria to a large extent.

By using more open technological and marketing strategies to increase the level of technological and market innovation in their product mix several companies in our previous research, such as Pharmacia and Perstorp, have historically been able to achieve successful industrial transformation by radical product innovation. Essentially this has meant going from a situation which is more low-tech to one which is more high-tech, for instance from basic chemicals for established uses to fine chemicals and chemical diagnosis in new applications. Similarly, many low-tech companies with problems concerning profit should be able to achieve successful product innovation and industrial transformation by employing more open development strategies.

Agroindustrial companies could, for instance, move in the direction of pharmaceutical companies, either by developing functional foods, that is food products with medical value (Mark-Herbert and Nyström, 2000) or by utilising their processes and raw materials for developing substances

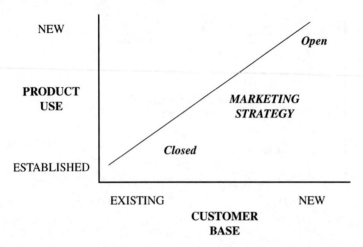

Figure 5.3 The main dimensions of more closed and more open marketing strategies

also for medical use. For example, Pharmacia finds the basic source for its world unique eye operation substance, Healon, in rooster combs. This is supplied by a farmer/entrepreneur who has helped to breed a suitable stock of roosters with giant combs. This entrepreneur also used eggs to produce antibodies for medical use and has been actively involved in research to develop this technology in a venture company.

A basic assumption in our framework is that it is necessary to consider both the company and product level – and the interaction between the two – if we are to understand how companies may develop radically new products. Open technological and marketing strategies are more concerned with the overall ability of companies to develop and market new products, than with the specific products that may result. At the same time the existing product mix is the baseline companies use to develop their core competencies (Prahalad and Hamel, 1990) and technological potential for finding new products and markets. In the following discussion we therefore will be concerned both with the company level and the product level of analysis.

Empirical results

The specific study reported on in this chapter deals with product development strategies in thirteen. Swedish biotechnical and food processing companies namely Cerealia, Juvel and Abdon (grain processing), SSA (sugar processing), VL and SL (grain and farm supply), Bionova, ABP, Lipid Teknic and Filium (bioengineering, biotechnology, etc.) and Binol (vegetable oils), SSP (starch processing) and Svalöf (plant breeding).

Based on personal interviews and company data, detailed information was collected in 1990 on the overall product development strategies of these companies and the new products developed, and marketed by these companies in recent years. Our empirical data makes it possible for us to divide our 13 companies into our two main categories, based on how closely their actual product development strategies correspond to our theoretical classification.

We begin with our first example of a company mainly pursuing Strategy A, Cerialia, which is part of SLR, the major Swedish group of farm cooperatives. Its primary development focus is on increasing cost efficiency in the Swedish Milling and Bakery Industry, using existing technologies. To increase the value of basic cereals Cerialia also carries out basic research in this area. It has its own research foundation, funding about 23 projects in 1989 and 50 projects in 1990. These projects are concerned with research into how to increase the value of cereals for food purposes. Cerialia has also funded a chair in Cereal Technology at Lund University.

Most of Cerialia's R&D cooperation is within the milling and bakery industry, but there are a few examples of inter-industry cooperation, for instance with Alfa Laval, Svenska Fläkt and SPP. In our terminology this means that the company has a relatively closed technological strategy, based to a large extent on isolated technology use (cereal technology) and internal orientation, (in-house research), but with some research cooperation and synergistic technology use. It wants, however, to move in the direction of Strategy B, by cooperating with universities and other companies within and outside of Sweden. Within the milling industry Cerialia, for instance, has R&D cooperation with a Swiss company, and in the bakery industry it is carrying out research together with English and German companies. Cerialia is also combining its knowledge of cereals with SPP's knowledge of starch, to try to improve the quality of its products. Cerialia's marketing strategy is highly closed, since almost all its new products are modifications of existing products, directed towards established customers

The second grain processing company, Juvel, is part of KF, the major Swedish food cooperative group. Its main objective is, as in the case of Cerialia, to concentrate on improving its basic assortment of milling and bakery products, directed to food retailers. Juvel's basic technology is processing grain from the grower to the retailer, using established know-how and little outside R&D cooperation. The company has its own laboratory for quality control, where most of its product development is carried out. Its main technological cooperation is with other consumer cooperatives in Sweden, to give and receive expertise on grain and pesticide analysis, and with outside customers, such as dog food companies, bakeries and baby food manufacturers to help them in their final processing of grain.

The company's goal is to concentrate its production by reducing its product line, rather than to diversify by developing new products. Its technological strategy therefore is quite closed – more so even than Cerialia's – and mainly characterised by isolated technology use and internal orientation to improve its existing products. Juvel does not carry out any basic research of its own and turns to universities, such as the Agricultural

University of Sweden, if they need help in this respect. Its marketing strategy is very closed, emphasising product modifications and existing customers.

Abdon, our third grain processing company, is a large privately owned international group of companies, which mainly produces basic cereals and bake-off products for the consumer and producer markets in Sweden and abroad. The company views milling as a very low-tech type of business, mainly producing low value-added standard products. It has very little research of its own or research cooperation, but it has carried out some nutritional research on oats in its American branch, to develop a new product, OatTrim. This is based on synergistic technology use, combining nutrition, food chemistry and microbiology. It is an ingredient for reducing fat in products such as ice cream and mayonnaise.

Most of the product development carried out in Abdon is, however, directed towards the internal development of existing technologies. This in order to improve the production of its basic ingredients for traditional products, such as bread and breakfast cereals. It therefore has a relatively closed technological strategy, emphasising internal orientation and isolated technology use. Its marketing strategy, as in the case of Cerialia and Juvel, is also relatively closed and mainly focused on product modifications and existing customers.

SSA mainly produces sweetening products for consumer and industrial uses and animal feed. Its main technologies are sugar processing and biotechnology and it carries out decentralised research in its product divisions and overall research in SSA Development. Historically, SSA has had a relatively open and innovative product development strategy, but from the late 1980s its technological and marketing strategies have radically changed. SSA has, since then, greatly reduced its research intensity, and moved away from basic research in sugar technology, towards more applied research.

At the same time, SSA has increased its marketing efforts for existing products and become a very marketing oriented company. Most R&D is carried out within the company to refine its existing technological base, the large scale processing and refining of sugar as a raw material for existing products. R&D cooperation is mainly with other companies in the sugar industry. SSA's technological strategy therefore has become quite closed, stressing internal orientation and isolated technology use, and its marketing strategy relatively closed too, emphasising existing products and customers.

VL, a regional grain and farm supply cooperative in Western Sweden, focuses on the distribution and processing of grain for animal feed and human consumption. Its technological strategy is relatively closed, based on internal orientation and isolated technology use. To become more innovative VL created a research foundation in 1986. The aim of this foundation is to carry out basic research within the technology areas of interest to the company, especially with regard to oats and oat fibres.

Based on the ideas originating in this foundation, external cooperation with outside researchers and companies will be used to develop new business proposals. Possible new products are yogurt, snacks, sandwich spreads and horse feed based on oats, and cooperation partners might be food companies or universities. Oat yogurt is an example of a new product based on synergistic technology use, combining oat and dairy technology. Other interesting research projects are the development of oat fat and oat starch, which have interesting food and medical properties. Swedish oats are also known to have a low protein content, which is desirable in feeding horses and could become an interesting export possibility.

VL's ambition is to achieve a more open technological strategy, based on external orientation and synergistic technology use. The company's original marketing strategy was also relatively closed – emphasising existing products and markets – but here again the company wanted to achieve a more open marketing strategy, to match its technological aspirations.

SL, a regional grain and farm cooperative company in southern Sweden, also has grain products for animal and human consumption as its main businesses. As in the case of VL, it has a relatively closed technological strategy and marketing strategy, but aspires towards more open strategies, by stressing cooperation with other companies, new technological combinations, and new markets. In 1985 it created a new venture department which will carry out basic and applied research, particularly with regard to oats and plant fibres.

One example of SL's external orientation and synergistic technology use in product development is combining plant technology with brewery technology in beer production, together with the Swedish brewery Pripps. Another is the industrial use of plant fibres, for instance in paper-making or for energy, where SL has been part of a joint research project with industry and universities. As in the case of VL, SL is trying to open up its technological and marketing strategies in order to become more innovative.

On the basis of our data, we have classified Cerialia, Juvel, Abdon and SSA as clear instances of companies using Strategy A, with SL and VL as more transitory cases, moving in the direction of Strategy B. We will now turn to the companies whose strategies may be classified as examples of Strategy B in our framework.

BioNova is a small bioengineering firm, working mainly in the area of enzymatic processing of grain. Its business idea is to separate grain into various components, fibres, carbohydrates and proteins, and market them for different applications. It has a relatively open technological strategy, based on synergistic technology use and external orientation. Its R&D cooperation is mainly with universities and companies in food processing and pharmaceuticals. It has a relatively open marketing strategy, searching for new markets for its products, for instance in bread mixes, vitamins and protein drinks for body building.

ABP is a small venture company, specialising in microbiology and biotechnology. This company was started in 1989 as a joint venture between several farm cooperatives in Sweden to develop new business ideas and carry out strategic R&D. In 1991 the company had 16 employees, among them microbiologists, biotechnical engineers and laboratory workers. This mixture facilitates development based on synergistic technology use and also external orientation, since many employees have kept up their contacts with their previous employers.

ABP engages in intensive R&D cooperation with universities and other companies and research workers coming from different disciplines. It is active in a number of research foundations, for instance in the area of membrane technology and biotechnology, and carries out research commissions for a number of companies. ADP therefore is a company with a very open technological strategy, based on external orientation and synergistic technology use and an open marketing strategy, based on selling its ideas to a wide range of companies in different technology areas. It also helps its parent companies to become more innovative by acting as a link to new knowledge and outside information.

Lipid Teknik is a small bioengineering company, which was established in 1986, producing and marketing special lipids to the pharmaceutical, food processing and cosmetic industry. It is a very innovative company where all employees are expected to take an active part in developing its technologies by publishing scientific articles, taking part in conferences and contacting potential customers. Basically the company has two types of projects, internal projects and commissioned projects. In both cases synergistic technology use is of dominating importance and this takes place in the interaction between the company's own and outside experts. Combining process engineering with lipid technology is an example of synergistic technology use, of major importance to Lipid in developing new products.

Lipid Teknik's technological strategy, thus, is highly open, stressing external orientation (for instance R&D cooperation with customers and joint university research) and synergistic technology use (for instance combining lipid technology with medical and food technology). It also has a highly open marketing strategy to find new customers and applications for special lipids.

Filium is a company in the rape seed oil processing business, specialising in industrial lubricants. It originated in 1970 as part of Karlshamn, a major Swedish company producing vegetable oil. The basic idea was to substitute vegetable oil for mineral oil as a cutting oil in the mechanical industry. This has clear health and environmental benefits and the company was a pioneer in developing this concept. To develop the idea Filium used a large number of external experts in mechanical engineering, health and environmental issues and also recruited outside experts to join the company.

Filium's technological strategy is quite open, combining biological and medical technologies in synergistic technology use and emphasising external cooperation with consultants and other companies. It has obtained patent protection for its applications, which indicates their high level of innovation, but its know-how is its strongest competitive strength. Filium's marketing strategy is relatively open with regard to finding new customers.

Binol is a company, which is also a spin-off from Karlshamn, developing and selling products based on processed rape seed oil.

Among its products are chain-saw oil, hydraulic oil and a penetration oil, to strengthen the effect of plant protection products. All these products are based on Swedish rape seed. Binol's product development resembles Filium's development and has its basis in the same basic competence, Karlshamn's raw material knowledge. This technological base was combined with technologies from outside the company, to provide specific applications to different end uses. Development partners were Karlshamn's R&D department, other companies and universities and research institutes in Sweden, Switzerland and the USA. Customers, for instance in the forest industry, were also used to further develop the products

Binol, thus, has a very open technological strategy, emphasising synergistic technology use (for instance combining lipid technology with microbiology) and external orientation (for instance company and customer cooperation). It also has an open marketing strategy, focusing on new products and market segments, such as vegetable based chain-saw oil for the forest industry and hydraulic oil for the food processing industry.

SSP is a farmer cooperative, starch processing company, developing and marketing its products and applications to the paper, forest, food processing and chemical industry. It is the company in our sample which has been most successful in going from low-tech to high-tech by employing open technological and marketing strategies. Thus it has been able to develop and successfully market a number of new products and develop new markets.

In 1980 the company was mainly providing the food industry with starch as a raw material, which is a good example of a concentrated, relatively low-tech strategy. In 1993 it had become much more high-tech and diversified by developing and marketing a number of technology intensive and differentiated products. Its industrial base was now much broader than food, and included paper, packaging, fish feed, wallpaper and ceiling material, health food and genetic engineering.

In 1980 the company devoted no specific resources to R&D. In 1983 it devoted one man-year; in 1990, 20; and in 1993, 21 man-years to research activities. It started a research foundation in 1983 with 10 million Swedish Crowns, which increased in 1990 to 32 million and in 1993 to 21 million Crowns. As a result it had entered into 2 joint research ventures in 1990 and 6 joint ventures in 1993. While it had achieved no patents from 1983 to 1990 it had more than 5 patents in 1990 and more than 10 in 1993. This clearly indicates that the company had been successful in its strategy for technological and market upgrading, to achieve more diversified and innovative new products.

SPPs technological strategy is relatively open, stressing external R&D orientation by cooperating with customers and universities and synergistic technology use, by combining for instance fibre technology with starch technology. Among the wide range of development partners are the Agricultural University of Sweden, Lunds University, The Caroline Hospital in Stockholm, Industrial Research Centres in Forestry and food and a number of large private and cooperative companies, e.g. Eka Nobel and Svalöf. SPP's marketing strategy is also very open, with the company actively searching for new applications and customers in a wide range of industries, from agriculture and forestry to food and medicine.

Our final example of a company which mainly uses Strategy C is Svalöf, a 50 per cent government and 50 per cent farm cooperatively owned plant breeding company. Svalöf has, for instance, used genetic engineering to obtain new breeds of potatoes, better suited for producing starch useful for applications in the paper industry. This has been done in cooperation with SPP in a jointly owned development company, Amylogen. Svalöf has also cooperated with SL, VL and Semper to develop oat varieties with higher fat content and other desirable qualities for medical use.

Its technological strategy, thus, is highly open, emphasising external orientation by R&D cooperation with industry and universities in Sweden and abroad and to some extent synergistic technology use, by combining plant breeding with industrial process technology, e.g. starch processing. Its marketing strategy is relatively open in the sense that the company has been successful in finding new customers for its technology, for instance SPP, VT and SL.

We thus see that Bionova, ABP, Lipid Teknik, Filium, Binol and SPP are all clear instances of companies following Strategy B, that is moving from more low-tech to more high-tech product development, by increasing the level of technological and market innovation in their new products. They also all have been quite successful in finding new markets for their new products.

In Figure 5.4 we have classified the thirteen companies with regard to how closed or open their technological and marketing strategies are, based on our interview data. We see that the companies most closely following Strategy A, Cerialia, Juvel, Abdon and SSA, all have relatively closed technological and marketing strategies. This is what we would expect from our theoretical framework and previous research, if they are to be successful in pursuing this strategy. If, however, they want to move in the direction of Strategy B, our research indicates that they need to open up their technological and marketing strategies, to increase their technological and market potential. VL and SL are evidently working in this direction, but it is still an open question to what extent they will succeed in their ambitions.

The companies classified as being relatively successful in implementing Strategy B – which means that they have already developed and marketed new products with a relatively high level of technological and marketing

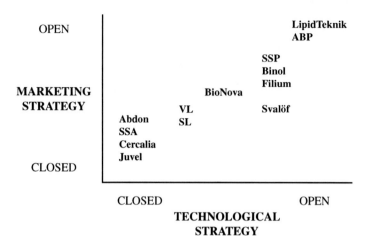

Figure 5.4 Classification of companies with regard to how open or closed their technological and marketing strategies are

innovation – all have relatively open technological and marketing strategies. This is particularly the case with regard to the most pronounced Venture Companies, Lipid teknik and ABP, but the other companies in this group also have relatively open strategies.

The main conclusion, therefore, is in line with our earlier studies. If companies want to move from a more low-tech to a more high-tech development situation, to find and successfully market radical new products, they need to be flexible and to open up their technological and marketing strategies to facilitate creative change. Both the need and potential for such change would appear to exist in many industries and companies today, not only in the agroindustry companies we have discussed above. At the same time our results indicate that this type of company should be well positioned to benefit from these possibilities by moving into new technology areas and markets such as bioengeneering, pharmaceuticals and health products.

Summary

This chapter shows the importance of strategic reflexivity and creative management to companies faced with the need to develop radically new products and services, to succeed in a rapidly changing technological and market environment. Open technological and marketing strategies, based on new technological combinations and changing customers needs, are necessary for companies to successfully move from a low-tech to a high-tech competitive position. The economic innovation literature, following Schumpeter, mainly focuses on the need for technological change. The management literature, on the other hand stresses the market as the main source for innovation.

In this chapter it is argued that we need to view the interaction between technological change and market opportunity as creating the main potential for successful innovation. Strategic reflexivity must be based on a flexible view of both technological and market possibilities. Innovation management must recognise and realise these possibilities and creatively guide and coordinate their development in a desirable direction.

6 Dependent innovation

Low technology and small enterprises in Denmark

Andrea Gallina

Introduction

The globalisation of industry which has taken place since the 1970s and has been driven by the revolution in the information technologies, the reduction of transport costs, the liberalisation of financial services and the de-regulation of the economy has produced a profound change in the organisation of industrial activities worldwide.[1]

The manufacturing sector is facing shorter product life cycle, increasing competition, and increasing markets opportunity. As a consequence firms compete increasingly on non-price factors such as innovation, quality, flexibility and specialisation (OECD 1996).

This has led to the disintegration of large companies into horizontally integrated units, and has produced a constellation of small and medium sized-enterprises (SMEs) linked to various extents in networks of producers. The common characteristic of these technological-productive *filieres*[2] that constitute networks is the presence of a leading company, which is depository of the knowledge responsible for the radical innovations and which is well integrated in the international markets. At the 'centre' of these networks there is generally a large enterprise, which carries out research and development (R&D), and is responsible for inventing new products and materials. Whereas at the 'periphery' there are the low R&D intensity sectors where SMEs are more important (traditional industries and niche markets), and innovation is incremental and based on the modification of existing products. Their level of internationalisation is low and generally pursued through export, mostly via intermediaries. In these sectors innovation is no longer an exclusive internal activity of firms, nor is it a mechanistic sequence from research to production and to the market, in which research is the driving force (Kaufmann and Tödling 2000: 30). Innovation can thus be considered as a non-linear and interactive process in which the external environment plays a substantial role (Dosi 1988), and in which non-R&D[3] activities must be taken into due account (Sterlacchini 1999: 819). North and Smallbone have reached analogous conclusions emphasising that in mature, craft based sectors, innovation in small firms is likely to take the form of design modification and incremental changes (North and Smallbone 2000: 147).

The analysis of the interaction between large and small firms and of the small firms' needs and learning capability in an open context can shed some light on the dynamic of the process of innovation. In this chapter the process of innovation is studied by looking at the interactions within associations of small and large firms, and between them and the environment, taking as a unit of analysis Danish firms, given the fact that Denmark is fairly open to the rest of the world and that SMEs are the prevailing form of industrial establishment.

Three cases enable us to understand the internal dynamics of innovation and the interplay with the external context. The SMEs interviewed in this study are in the metal, printing and packaging industries. They have been selected according to their size (between 49 and 100 employees) and for being suppliers of large companies in highly internationalised sectors such as food, pharmaceutical and metal manufactured products.

According to the OECD classification, the sectors represented in the case study are defined as being of low technology.[4] The importance of qualitative in-depth analysis for the measurement of innovative activities in low-technology sectors and SMEs has already been argued (cf. Hansen and Serin 1997). Following this method semi-structured interviews were conducted for an average of two hours with either the production manager or with the owner of the firm during the months of October and November 1999.

The remainder of the chapter is organised as follows. In the second section the concept of innovation as strategic reflexivity is discussed in the context of small and medium-sized businesses in the manufacturing sector. It is stressed that the strategic reflexivity of the process of innovation is highly dependent on the type of relationship between the firms and between them and the market. Next, we discuss the role of SMEs in Denmark and describe the internationalisation of the branches to which the firms selected for this study belong. Despite the lack of homogeneous statistics, this section attempts to describe the structure of the chosen sector and their level of internationalisation. We then go on to discuss the cases. It is emphasised how the interaction between the internal and external elements in the firm leads to innovative activities. The process of innovation in those SMEs which supply large firms is highly dependent on the knowledge transferred by the large enterprise, and it is the result of the transformation and mobilisation of the internal assets of the small firm. The conclusions sum up the discussion and put forward some hypotheses that can have important implications for future research.

Innovation as strategic reflexivity

The chapter attempts to contribute to the understanding of the process of innovation as strategic reflexivity. Given that production is globalised and increasingly intertwined in networks of producers, the source of advantages are to be found in the collaboration that firms have with each other. Innovation is thus regarded as the result of the interaction between factors internal and external to the firm. External sources of knowledge include the customers, the international trade fairs, the suppliers and the institutions such as the technological institutes. Internally the firms rely on a set of 'assets' such as capital equipment and the combination of formal and tacit knowledge embedded in the organisation (Polany 1962). The evolution and combination of these assets are particularly relevant for the firm's survival. In this perspective the processes can be perceived as strategic and reflexive (cf. Sundbo and Fuglsang in this book).

Innovation in SMEs in the low-tech branches is *reflexive* because it involves a systemic consideration of the expectations of other actors, i.e. the customers and the suppliers at the various levels of the *filiere*. SMEs are part of the larger systems of production and, whether they reach the final consumer or not, they interact constantly with other actors and the market. In this sense, innovation can no longer be considered as a hierarchical process but as an interactive one. Specialisation, segmentation of markets, and internationalisation stimulate the level of interactions between agents and thus raise the level of conflicts and expectations.

As a consequence, innovative activities increasingly reflect the complexity of the environment, but in the specific case of SMEs in small open economies the solution of conflicts and the satisfaction of expectations through innovation could be more dependent on the external context and hence should be negotiated. In the low-tech manufacturing sector, SMEs participate in negotiations from a weaker position. Even though innovation can lead to specialisation, and can confer upon the firm a clear role in the international division of labour, the source of relevant knowledge is in the hands of the large multinationals.

From this point of view the innovative activities are less *strategic*. Still, the firm's stock of knowledge, embedded in the hands and heads of the workers, represents an important asset the role of which in the negotiation and de-codification of the knowledge transferred by the large customer deserves to be highlighted.

Therefore, innovation is 'strategically reflexive' if it enables the firm to be an active part of the interacting network in which expectations are created and goals are achieved despite the few resources available.

The application of the concept of innovation as *strategic reflexivity* depends very much on the level of independence firms and sectors have from other firms and sectors. The strategic reflexivity of innovation in SMEs can be compared to a process of building up, through interactive learning, intangible assets from the main block of knowledge supplied by the big firms. This type of innovation can also help the firm to find alternative roles in the highly uncertain environments dominated by large and footloose enterprises.

In the case study the ability to create expectations becomes central for the survival of the firm and this can be achieved through the firm's rapid adaptability of both process and product to the needs of the customers. Strategic reflexivity is seen in the ability of the SMEs to acquire, absorb and personalise the knowledge produced in interaction with the large firm, and to influence the decision of the big company in adapting and modifying the new product to the capability of the technology and knowledge available.

Small enterprises in the process of globalisation

When the industrial structure is characterised by the presence of SMEs and by low R&D intensity sectors, the process of innovation has important consequences on the survival, growth and reproduction of firms as much as in high technology and sectors dominated by large firms. SME's importance in the process of economic growth of industry has been acknowledged in the literature (Piore and Sabel 1984; Sabel and Zeitlin 1997; Becattini 1989; Sengenberger 1989).

Despite the end of mass production, price and economies of scale are still very important, but competition is also based on factors such as innovation and the ability to adapt to customer needs. These factors have always been claimed to be the source of advantage that enable SMEs to remain in the markets, while providing stability to the productive system in which they are embedded (Rothwell and Zegveld 1982; Rothwell 1984; Acs and Audretsch 1990a and 1990b; OECD 1997; Pratten 1991).

The role of SMEs in the new division of labour can be analysed from this perspective. SMEs are facing increasing pressure from the processes of de-centralisaton and de-territorialisation of production. The growing demands for special products favour SMEs and partially diminish the importance of cost minimisation and mass production. Parallel to this, the restructuring of large companies in small and horizontally integrated flexible units and the de-centralisation of the activities to networks of dependent subcontractors have reduced the traditional advantages of SMEs and weakened their negotiating power. Competitive advantages, especially on the global markets, have to be found in other fields.

The changing business conditions and the uncertainty of the environment place a greater pressure on their activity than in the past. SMEs' role as suppliers of large firms that are connected to the global market increases their exposure to the volatility of international prices and the demand of more sophisticated markets.

In this perspective, even firms without any registered international activity, can be connected to the global markets while supplying large internationalised enterprises. Large enterprises are those carrying out the basic research activities and defining the products' characteristics in terms of design and material.[5] The final result of these activities is a very specific demand to the supplier, which is generally a SME.[6] Competition to remain in the network or association of suppliers is hard. The SMEs that innovate survive this first selection process and their capabilities are 'upgraded' through the transfer by the large firm of stocks of knowledge about new products, process or material characteristics.

Firms receive constant feedback during the production activities that can be transformed in idiosyncratic knowledge leading to relevant changes. Thus, interactive learning[7] and networking as a strategy for finding sources of knowledge are particularly important.

The internationalisation of the economy puts the survival of SMEs at risk. The traditional advantages of SMEs of being able to deliver small batches to local markets and provide the customers with specific products are swept away by the spread of these new forms of organisation and production leading to mass customisation. Industries in which the presence of SMEs is extensive are increasingly trade oriented, and SMEs face greater import competition in their home markets and export competition as they extend their operations to foreign markets (OECD 1996: 53). Furthermore, large firms' rationalisation strategies are dramatically reducing the number of suppliers, therefore to remain in the association of suppliers became an issue of survival.

This also means that the collaboration agreements between firms and the creation of networks are increasingly strategic. The establishment of backward and forward linkages enabling interaction, learning, and the continuous transfer of knowledge contributes to the upgrading of firms' capability to innovate. The paradox of the Italian industrial districts in which innovation and growth have been possible in low-tech and traditional industries has led to re-considering the sources of advantages in those tacit, uncountable, and hidden mechanisms that produce the relevant knowledge for success (Bagella and Becchetti 2000).

Also in isolated firms, i.e. those not located into a specialised cluster of industries enjoying external economies, the opportunities and constraints deriving from increasing internationalisation play a substantial role. It is also necessary to stress that the implications of a change in the environment and the innovative activities that eventually follow are different and contextual to the firm and the sector in which they are operating.

Dependent innovation in low-tech sectors

Due to the constraints that block SMEs in the low-technology sectors from undertaking activities leading to radical process and product innovations the interaction between different sources of knowledge is central. These sources are localised both within the firm, such as the skills and competences[8] of the workers and the entrepreneur, and outside the firm, such as the customers' and suppliers' knowledge, and the support institutions (technology centres, certification centres, etc.).

In a context such as Denmark where there is a high degree of commercial openness and internationalisation of manufacturing activities, the processes of learning and innovation have a trajectory that is strongly influenced by external elements. The enterprises receive incessant feedback from the environment that can be traced in both organisational learning[9] and in product and process innovation. The internal resources of the firm are mobilised to support the transfer, absorption and adaptation of the knowledge exchanged. During the interaction process new competences are created and the stock of idiosyncratic knowledge increased.

The type of innovation depends very much on the type of sector. The firms surveyed are 'supplier dominated' or 'specialised supplier' (Pavitt 1994). In these firms product and process innovations needed for their survival and growth do not necessarily require R&D since, on the one hand, product innovations are mainly of incremental nature and process innovations depend, essentially, on the supply of (and interaction with) capital goods industries (Sterlacchini 1999: 830). In both product and process innovations interaction and learning create new competences and knowledge that lead to innovation. The exchange and accumulation of knowledge are the central elements in the process of innovation. In our case this dynamic is dependent on the power structure of the relationship and is influenced by the evolution of the external environment (Figure 6.1).

Large firms retain the relevant knowledge about processes, materials, products, design, and also of the outlets. Small firms have the knowledge about the production processes. The combination of the two stocks of knowledge will lead to the creation of an entirely new product or the modification of an existing one.

From the figure it evinces that the flows of knowledge are greater in firms organised in specialised clusters. However, the firms in the sample are not enjoying these kinds of externalities. They produce the entire product in the company, and with the exception of the relationship with their suppliers, the main knowledge source is the large company. What is important for our analysis is to highlight that despite the flow being bi-directional, the one originating from the large enterprise is thicker than that originating from small firms. In both cases 'thickness' increases through the exchange, but the relevant knowledge from which a new product originates comes from the large firm's R&D activities.

This has important implications for industrial and innovation policies. They have to be preceded by an understanding of the transfer of knowledge mechanisms before appropriate development frameworks can be formulated. This process involves an analysis of the will of the parties to transfer knowledge and the capacity of the recipient to absorb and transform it. But also, the proper functioning of the communication channels (both formal and informal), the mutuality of the learning process and the possibility of de-codifying the embedded knowledge (cf. Whiston 1996). The importance of the institutions is therefore central in setting the conditions

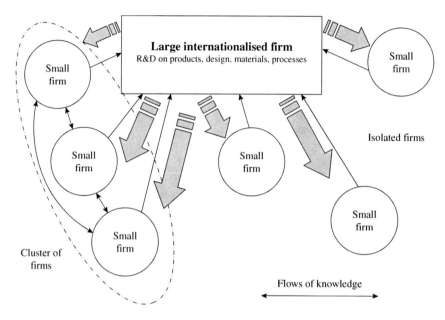

Figure 6.1 Knowledge flows in the manufacturing sector

for the successful creation and exchange of knowledge. The success of the transfer depends on the terms of reference established by the institutional channels and on the possibility given to the local agents of receiving and absorbing the knowledge available.

The Danish context

In Denmark SMEs remain the prevalent type of firms and the industrial structure has not changed despite some major mergers. SMEs are considered to be those that employ less than 100 employees.[10] They represent about 95 per cent of the total number of workplaces, account for 40 per cent of the employment in the manufacturing industry and 25 per cent of private sector employment. The contribution of SMEs to the internationalisation of the Danish economy is mainly through exports (see Table 6.1).

SMEs are concentrated in the traditional sectors characterised by the use of 'low technology', and are often family-based. The owner, chairman of the board and managing director are often the same person; whereas to meet the legal requirements, other members of the family become directors, 'the old aunts and uncles boards of directors' (Madsen 1986: 4). They are oriented towards both the local and international markets, have a high level of specialisation, and mostly occupy market niches.

Table 6.1 The role of SMEs in the Danish
manufacturing sector (1997)

Size	Enterprises	Employment	Export
<10	60.9	10.0	3.6
10–19	15.9	7.6	2.2
20–49	12.8	12.4	6.5
50–99	4.9	10.3	13.5
100+	5.5	59.7	74.2
Total	100.0	100.0	100.0

Source: Elaboration from Denmark Statistics, various years.

The absence of a large base of big enterprises and the high level of internationalisation of the manufacturing industries has had two specific effects. It contributes, first, to the creation of long and extended trade and production networks and, second, to the pulverisation of productive activities into a myriad of small dependent subcontractors. This generally places the SMEs in a less autonomous position with respect to the process of innovation and highly dependent on fluctuations of the international markets and multinationals' strategies.

Openness and internationalisation of Danish manufacturing

The Danish economy is relatively open to the rest of the world with a trade balance showing a slight surplus. About half of the production from the manufacturing industry is exported.[11] Since the early 1990s, the most profitable sectors for the Danish economy have been those of machine and transport equipment and food and live animals, which taken together contributed for about 50 per cent of total trade.

The branches of the manufacturing sector that have registered a positive balance are furniture, sanitary and heating equipment, wood and cork products, glass products, precision instruments and machinery for specialised industry (see Figures 6.2 and 6.3).[12]

Since the early 1990s the rate of openness[13] with the rest of the world has increased constantly. The openness is due to a dependency on imports in almost every industry in the manufacturing sector, with the exception of the sectors in which the country has a high degree of specialisation (medicine and pharmaceuticals, general industrial machinery, precision instruments, furniture, non-road transport equipment).

The firms studied belong to branches whose rate of openness is relatively low. This can be due to the fact that it is a production with a high volume/price ratio and therefore less interesting from an export point of view. Nonetheless, these branches are producing for the most open sectors, such as packaging and printing for the pharmaceuticals and food industries and

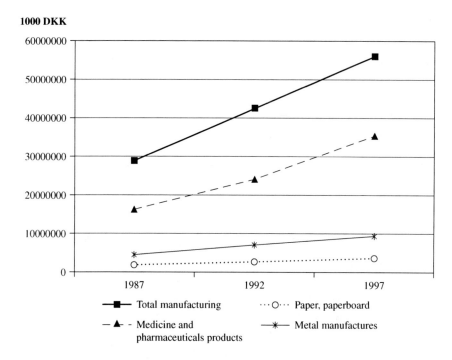

1000 DKK

Figure 6.2 Evolution of manufacturing exports and in selected industries, 1987–92, in 1000 DKK

metal products for the specialised machine and metal manufactured industries. The areas of the world economy to which they mostly supply has also increased constantly (Figures 6.3 and 6.4). The openness of the economy, linked to a dynamic and growing export and import puts SMEs in a very open context in which information about new markets, products, process and organisation flow abundantly.[14]

The case study

Three case studies present the process of innovation as the outcome of a circular interplay between internal and external sources of knowledge. The firms have been selected according to the size (between 49–100 employees) and the sector (low-technology) and on the basis of supplying a large company, although not linked in a subcontracting relationship. The firms are located in different regions of the country, close to large-middle towns and situated within industrial areas.[15] All of them are specialised in the supply of a market niche. Their range and volume of products are limited, but of good quality (see Table 6.2).

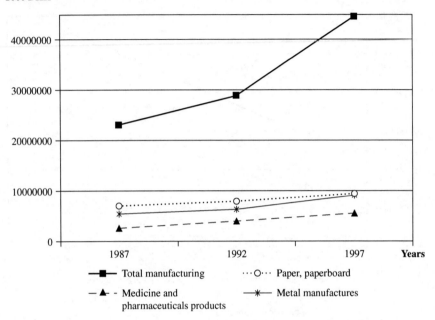

Figure 6.3 Evolution of total manufacturing imports in selected industries, 1987–97, in 1000 **DKK**

The firms interviewed have different degrees of exports and relationship with larger companies. Firm α is a manufacturer of aluminium products for different uses and for different industries. The production ranges from components for hydraulic pumps to supports for garden lamps, with a high degree of internationalisation (70–80 per cent of the production is exported) in a dynamic and expanding market. The firm supplies mainly a large Swedish company whose production is sold in Europe and Latin America. Part of the production is also exported to Germany through a Danish intermediary. The rest of the production (20–30 per cent) is sold on the national market that despite the low number of competitors cannot absorb more than that. The company does not possess an R&D department. The customer provides the design of the components and of the technical specifications. New products are developed according to the specific request of the customer.

Firm β, is a manufacturer of complete packages (carton, labels, leaflets, compact label and pressed folio) for the pharmaceutical industry. At the time of the interview the company had just merged with a Swedish medium-sized enterprise. Although it had a consistent growth over the last few years, in order to overtake the financial obstacles to the purchase of

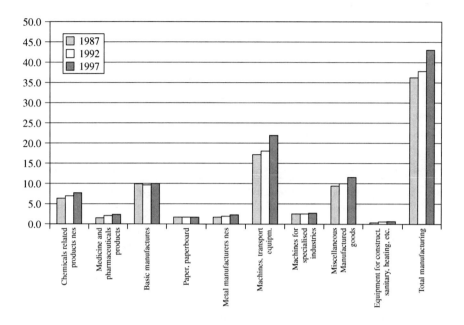

Figure 6.4 Sectorial rate of openness

new technology and to have a better bargaining with the suppliers of raw materials, the company felt the need to merge with another company. The merger has not had a substantial impact on the independence of management and production and it has instead reinforced the specialisation of the two companies. The Danish firm is still making carton packaging while the Swedish partner produces plastic containers and the previous carton production has been transferred to Denmark. Innovation in products depends largely on the high specificity required by the sector supplied. The large pharmaceutical or food companies provide the materials and design requirements each time a new package is needed.

Firm γ, is also a folded carton packaging producer, it has almost no exports but supplies large companies in the pharmaceuticals (Novo Nordisk) and food industries (Toms' Chokolade) fully integrated into the world economy. In the same specific sector competition is very hard due to the presence of at least ten other companies of similar size and by the entrance in the sector of the flexible packages made of plastic. Also the entrance in the market of neighbouring countries with lower labour cost, such as Poland is seen as a threat. Product innovation is the result of interaction with the customers that provides the design and the material minimum requirements.

Table 6.2 Schematic description of the companies interviewed

	Type of production	Type of market	Type of customer	Type of technology	Type of innovations
Case α	Components in aluminium for hydraulic underwater pumps and for furniture industries	Mainly international (70% of exports), directly or through an intermediary	Supply especially to a large Swedish company with customers all over the world	Low, but with the presence of very advanced robotic systems just to perform a specific operation	Incremental product innovation
Case β	Printing and complete packaging for pharmaceutical industry	Scandinavian, Germany and Poland	Pharmaceutical MNCs	Low but flexible with the use of CAD. The pre-press operation is computerised	Incremental product innovation
Case γ	Folded-carton packaging for food and pharmaceutical industries	Very limited exports, studying the possibility for expanding them in the Baltic countries	Supply to either large multinationals' subsidiaries or companies with a high degree of inter-nationalisation	Low, but with a combination of new and old high-quality equipment	Incremental product innovation

Process innovation is very costly and investments in new machines occur every few years or if there is a specific reason, as for example, to acquire a big customer whose product needs a specific operation. In this case economies of scale justify the investment. It is, however, felt that asking if the company has introduced new machinery in a given period can produce a shortcoming in the research. In fact, the technological level of the firm's capital stock should be taken into account. As pointed out by Sterlacchini, in the case of a discontinuous pace of technological change, characterised by periods of relative technological stability in which only small improvements are made to production processes, it is not necessary for a firm to replace its capital stock through intense acquisition of new machinery during these periods (1999: 823). This is particularly true of low-tech industries. In these SMEs significant process innovations (such as CAD-CAM) were already introduced in the mid-1980s.

From the case studies it emerges that product innovation is strongly dependent on the development of the customer's technical knowledge. The large company delivers the drawing with the specifications and with the type of material to be used. The choice of the material is particularly relevant to the pharmaceutical and food industries.

Innovations are therefore incremental, continuous and mainly encompass the modification of existing products. We are in a situation that reflects the incremental pole of Abernathy and Utterback's dichotomy.[16] As pointed out by Hansen and Serin the people in the firm do not perceive this as innovation development but as design and customer adaptation of the product (Hansen and Serin 1997: 186). The circular link is established when the small firm sends back the prototype that can be closer to what the large firm was expecting and then the interplay starts. The skills and the flexibility in satisfying the customers' requests and perhaps modifying them according to the capacity and knowledge about the production process and products characteristics embedded in the firm are the main assets of these SMEs.

Incremental continuous innovation is the result of strong interactions within the company and with external sources both in case of intermediate products and finished products. The development of a new product is a process of 'rationalisation' of customer needs and supplier and knowledge. From one side there is a large company transferring knowledge about the product design and material and on the other there is the supplier that transfers the knowledge on the actual feasibility of producing the goods with a given technology and material. Both supplier and customer are recipient and transferor of knowledge.

Innovation springs from a long gestation period at the levels of the workshop and sales and production departments and in the meeting between the two companies' technicians. In the SMEs, the gestation period is the phase during which the information and the knowledge provided by the customer is absorbed, de-codified and transformed into new knowledge. The skill of the engineers and their contact with the workshop are central for providing the customer with the product required. But the actual production needs a continuous modification and adaptation of the design to the tasks the machines and the workers' hands are able to perform. This implies that the skill of the engineers and the ability of the workers (the intangible factors) enable the achievement of efficiency utilising machines that do not contain the latest technology.

Innovation is both a hidden process and an activity which is difficult to measure. Innovative activity is the outcome of both the internal interaction between production managers, foremen and sales managers and between them and the customer. Still the definition of the characteristics of a product and how they should be modified are dependent upon the power relationship which is established. The capability to accumulate this knowledge and have a shorter gestation period gives the firm the chance to respond quickly to old and new customers and offer a high quality and personalised product. The interactions with large and internationalised companies increase the opportunities of creating new knowledge, absorbing information originated for and by other settings. The source of varieties is enlarged and the risk, often faced by small firms, to be locked in due to lack of information from outside is reduced enormously.[17]

In this way SMEs continuously develop assets that have mainly an intangible nature, such as the know-how in the hands and heads of workers and entrepreneurs enabling them to absorb and diffuse the knowledge internally.[18]

Conclusions and implications for the process of innovation

The instability and uncertainty deriving from the process of internationalisation and the situation of dependency from large companies put SMEs' survival at stake. Their connection to global networks can be regarded as a source of advantage only if there is an active participation in the process of learning. In the sector analysed suppliers of raw material do not represent an important source of innovation and the main pushing factors come from the customers. The interplay between the customer and the SMEs leads to the upgrading of the firms' competences and scientific and technological capabilities. In other words, the small firm, although dependent on the knowledge and the economic power of the multinational, learns to solve new problems and to improve its production ability. It should also be taken into account that customers are global and satisfy international and differentiated markets. The firm is therefore connected to global networks and the knowledge made available is increasingly less rooted in the local/national system.

The search for 'competitive advantages' easily overcomes the 'cultural barriers', and cultural distance between agents and is often eliminated by interaction and trust building. The reduction of the fear of disloyalty diminishes as far as the business relationships and the international commitment of the firm become stronger.

The process of adaptation of the firm to new demands and new technological opportunities does not necessarily lead to radical products or process innovations. In the cases analysed, this applies to the sectors in general and where they are located; innovation is the output of a continuous process of adaptation and modification of product design following either random intuition at the workshop level, but especially interaction with customers. For these firms, and particularly for smaller ones and in low-tech sectors, the main step was to acquire computer aided and numerically controlled machinery. Once they have purchased that type of equipment they can keep the same level of technology for many years and therefore base their survival and growth strategy on other factors. Design and product adaptation become the central elements. Innovation is possible only with close co-operation and interaction among workers at the workshop level, between manufacturer and customers, between manufacturers and intermediaries and between competitors. In this way the stock of knowledge, the competence and the capability of a firm increase and sales are secured.

These final considerations represent a point of departure for the further development of a theoretical model of learning and innovation that considers each different level at which productive activities take place: local, national and global, and internal and external. How these different levels interact and how such approach supports policies having a local impact considering the global dynamics can represent the basis for further analysis.

From the results of the case study, it can also be concluded that innovation is increasingly interactive and based on the validity of intangible resources. But it is also dependent on the possibility of relying upon the knowledge produced and transferred to them by the large customer. The high level of specialisation of the firms analysed and their concentration on satisfying mainly one large customer's expectations are the central factor to the formulation of strategic reflexivity. Innovation is reflexive but the possibility of manipulating these expectations is limited by the power structure of the relationship in favour of larger firms. From the study it emerges also that innovation can be strategic if the mobilisation of the internal intangible assets can help to find alternative roles in the division of labour through specialisation.

The conditions that characterise the manufacturing sector today put the firms in a situation where they have to take into account the dynamics of many phenomena. Satisfying and creating expectations and negotiating a way of achieving them through continuous interactions is the objective of innovative activities. In this sense the strategic reflexivity of SMEs is a process dependent on the external environment but it is to a large extent influenced by the development of internal intangible assets.

Finally it should be stressed that a view focusing only on the internationalisation of the economy and technology as a process that contributes to increasing the sources of knowledge available and therefore innovation, underestimates its effects on the local production systems. Increasing decentralisation of production, the pulverisation of manufacturing activities and the uncertainty of the environment that is produced are giving SMEs a different role in the economy. The historical function of SMEs of providing goods for local markets and needs is now substituted by the involvement in footloose networks and the association of suppliers. As a result, the dependency of SMEs from large companies is increased and so is that of innovation activities. Whether this has an implication on the creativity and originality of SMEs' entrepreneurs is not easy to say, but still the impact of 'dependent innovation' on the socio-economic development of the territory in which they are located needs further investigation. From this point of view strategic reflexivity can be a conceptualisation that should also include the role of the expectations of the socio-economic and institutional environment in which innovation is taking place.

Notes

1 The term globalisation is used, according to the OECD, to indicate 'the trans-border operations of firms undertaken to organise their development, production, sourcing and financing activities. A distinctive feature of globalisation is the division of firm operation into separate segments carried out in different countries. The most prominent features of globalisation are foreign direct investment, various aspects of international trade, and international inter-firm collaboration' (OECD 1996: 15).

2 A *filiere* is the sequence of operations carried out in the process of transformation of raw material in a finished product. Each *filiere* is made of productive cycles and each cycle is divided in phases, each phase in operations and each operation in a number of basic movements. The diffusion of flexible mode of production and specialisation has increased the complexity of the filiere. Phases and operations are increasingly distributed among productive units.

3 Non-R&D activities are for example design work, engineering and pre-production development.

4 According to the OECD low-tech is an industry in which the gross R&D expenditure is less than 1 per cent of the turnover: OECD 1995, Paris.

5 Material characteristics are particularly important in sectors such as pharmaceuticals and food due to the high quality and sanitary standards required for these products.

6 Suppliers are generally more than one in order to avoid a bottleneck in case of problems in the supply chain of the requested product. But, with the new management strategies of large enterprises and the consequent rationalisation of production the number of suppliers is reduced.

7 The role of interactive learning as a driver of innovations has been deeply debated in the literature on the National Innovation Systems (Lundvall 1992).

8 On the concept of competence and the internal resources of the firm cf. *inter alia* Prahalad and Hamel and N. Foss. These elaborations draw from the work of Edith Penrose.

9 For the impact of internationalisation on the organisational structure cf. Nieminen and Törnross, 1997.

10 In general it should be emphasised that statistics about SMEs in Denmark are very inconsistent and are not homogeneous.

11 Denmark Key Figures 1999, Confederation of Danish Industries, Copenhagen.

12 A more detailed analysis of the pattern of specialisation could be carried out in order to highlight if the specialisation is linked to products that are increasingly relevant for the world market.

13 The rate of openness has been calculated in the following way: $(M+X)/MVA*1/2*100$, where M and X are respectively the total imports and exports in the manufacturing sector, and MVA is the manufacturing sector total value added. In case of surplus or equilibrium of the trade balance the rate of openness measures the share of value added generated by the sector producing export goods; in case of deficit of trade balance the rate of openness measures the dependency of the country from import. Anyhow, an increasing value of the rate of openness indicates an increasing openness of the economy analysed.

14 As pointed out by the OECD: 'the global expansion of industry has more general implications for small and medium-sized firms operating in internationally trading industries even if they have no international operation themselves. They are major suppliers to large firms, and there are rising competitive pressures on large firms to increase efficiency and purchase more goods, intermediates and services inputs externally. This provides opportunities for small firms to form subcontracting and supply linkages into large ones. But these linkages are increasingly

driven by international strategies, and local small firms face competition in supply from small firms in other countries, and from larger specialised international suppliers with internal R&D capabilities which smaller firms lack' (OECD 1996: 53).

15 Contrary to the general idea of 'natural' processes of localised learning as a consequence of spatial proximity, the firms interviewed do not have any particular benefit from being surrounded by other firms producing similar goods.

16 For a discussion of this approach to product development see O'Shea and McBain (1999).

17 In the specific case of small countries other external sources of knowledge are also very important. Through participation in international trade fair knowledge is produced and specific competences and capabilities are built. If the information about a new machine or about a new customer or product is considered relevant then the process will continue and the bits of information will be transformed into knowledge through meetings with the producers and the practical use of the equipment. The importance of the market for the acquisition of new knoweldge is highlighted also by Maskell (1996): 19).

18 On the absorptive capacity of the firm cf. the discussion in Mangematin and Nesta, 1999.

7 When market research may hinder the development of discontinuous new products

Paul Trott

Introduction

In his award winning 'business book of the year'[1] Clayton Christensen (1997) investigated why well run companies that were admired by many failed to stay on top of their industry. His research showed that in the cases of well managed firms such as Digital, IBM, Apple and Xerox, 'good management' was the most powerful reason why they failed to remain market leaders (*sic*). It was precisely because these firms listened to their customers and provided more and better products of the sort they wanted that they lost their position of leadership. He argues that there are times when it is right not to listen to customers. Recent research by Ovans (1998) supports this claim. He suggests that purchase-intention surveys are not effective predicators of sales of new products. The research revealed that people aren't generally reliable predictors of their own long-term purchasing behaviour. The type of question used and whether or not the question is placed in context greatly affects the reliability of such market research. James Dyson has good reason to be suspicious of the role of market research in new product development. Not only did he struggle for many years to get anyone in the UK to believe it was worth manufacturing his bagless vacuum cleaner, he faced the same scepticism when he launched in the US (Thrift, 1997).

Many industry analysts and business consultants are now arguing that the devotion to focus groups and market research has gone too far (Christensen, 1997; Martin, 1995; Francis, 1994). Indeed, the traditional new product development (NPD) process of market research, segmentation, competitive analysis and forecasting, prior to passing the resultant information to the research and development (R&D) department, leads to commonality and bland new products. This is largely because the process constrains rather than facilitates innovative thinking and creativity. Furthermore, and more alarming, is that these techniques are well known and used by virtually all companies operating in consumer markets. In many of these markets the effect is an over-emphasis on minor product modifications and on competition that tends to focus on price. Indeed, critics of the market-orientated approach to new product development argue that the traditional marketing activities of branding, advertising, positioning, market research and consumer research act as an expensive obstacle course to product development rather than facilitating the development of new product ideas.

For many large multi-product companies it seems the use of market research is based upon accepted practice in addition to being an insurance policy. Many large companies are not short of new product ideas – the problem lies in deciding in which ones to invest substantial sums of money (Trott *et al.*, 1995), and then justifying this decision to senior managers. Against this background one can see why market research is so frequently used without hesitation, as decisions can be justified and defended. Small companies in general, and small single product companies in particular, are in a different situation. Very often new product ideas are scarce; hence, such companies frequently support ideas based upon their intuition and personal knowledge of the product. This is clearly the situation with James Dyson's bagless vacuum cleaner (Dyson, 1998).

Morone's (1993) study of successful US product innovations suggests that success was achieved through a combination of discontinuous product innovations[2] and incremental improvements. Indeed, Lynn *et al.* (1997) argue that: in competitive, technology-intensive industries success is achieved with discontinuous product innovations through the creation of entirely new products and businesses, whereas product line extensions and incremental improvements are necessary for maintaining leadership. This, however, is only after leadership has been established through a discontinuous product innovation. This may appear to be at variance with accepted thinking that Japan secured success in the 1980s through copying and improving US and European technology. This argument is difficult to sustain on close examination of the evidence. The most successful Japanese firms have also been leaders in research and development. Furthermore, as Cohen and Levinthal have continually argued (1990, 1994) access to technology is dependent on one's understanding of that technology.

Adopting a technology push[3] approach to product innovations can allow a company to target and control premium market segments, establish its technology as the industry standard, build a favourable market reputation, determine the industry's future evolution, and achieve high profits. It can become the centrepiece in a company's strategy for market leadership. It is, however, costly and risky. Such an approach requires a company to develop and commercialise an emerging technology in pursuit of growth and profits. To be successful, a company needs to ensure its technology is at the heart of its competitive strategy. Merck, Microsoft and Dyson have created competitive advantage by offering unique products, lower costs or both by making technology the focal point in their strategies. These companies have understood the role of technology in differentiating their products in the marketplace. They have used their respective technologies to offer a distinct bundle of products, services and price ranges that have appealed to different market segments. Such products revolutionise product categories or define new categories, such as Hewlett-Packard's laser-jet printers and Apple's (then IBM) personal computer. These products shift market structures, require consumer learning and induce behaviour changes, hence, the difficulties for consumers when they are asked to pass judgement.

It seems the dilemma facing companies is twofold:

* *At the policy level*: to what extent should companies pursue a strategy of providing more room for technology development of products and less for market research that will *surely* increase the likelihood of failure, but will also increase the chance of a major innovative product.
* *At the operational level*: to what extent should product and brand managers make decisions based upon market research findings.

Contribution to the innovation management framework

Joseph Schumpeter's (1939) recognition of the importance of technology and the role of the entrepreneur for economic growth remains significant today in how we view innovation. It is the cornerstone of accepted theories of innovation. Indeed, later explanations and theories (cf. Arrow, 1969; Cohen and Levinthal, 1990; Dosi, 1982; Nelson and Winter, 1982) have built on Schumpeter's work and emphasised the importance of technology management and research and development (R&D) as key influences on innovation. In today's so-called knowledge-based economy, however, manufacturing is no longer viewed as the dominant industrial activity – marketing, promotion, branding and service have become the dominant industrial activities, and nowhere is this more apparent than in the e-commerce revolution. To what extent, however, is this justified? Is there now too much emphasis on these marketing activities? Market research results frequently produce negative reactions to discontinuous new products (innovative products) that later become profitable for the innovating company. Famous examples such as the fax machine, the VCR and James Dyson's bagless vacuum cleaner are often cited to support this view. Despite this, companies continue to seek the views of consumers on their new product ideas. The debate about the use of market research in the development of new products is long-standing and controversial.

This chapter addresses one of the key influences on innovation, that being the market. Clearly the market is where firms and products in particular compete; and firms must consider their customers and potential customers if they wish to compete. It questions, however, whether too much traditionalism in firms' interpretation of market possibilities can hinder the development of discontinuous new products (Hamel and Prahalad, 1994; Christensen, 1997). It is the responsibility of management within firms to provide room for 'scientific freedom' and R&D management in particular to facilitate controlled chaos. For, to shackle the creative minds of scientists and technologists to the myopic views of the market would be folly indeed, and would conflict with the fundamental requirements for innovation.

The use of market research in new product development: a review of recent thinking

In one of the most comprehensive reviews of the literature on product development Brown and Eisenhardt (1995) develop a model of factors affecting the success of product development. This model highlights the distinction between process performance and product effectiveness and the importance of agents, including team members, project leaders, senior management, customers, and suppliers, whose behaviour affects these outcomes. The issue of whether customers can hinder the product development process is not, however, discussed.

It is argued by many from within the market research industry that only extensive market research can help to avoid large scale losses such as those experienced by RCA with its Videodisc, Procter & Gamble with its Pringles and General Motors with its rotary engine (*Business Week*, 1993). Sceptics may point to the issue of vested interests in the industry, and that it is merely promoting itself. It is, however, widely accepted that most new products fail in the market because consumer needs and wants are not satisfied. Study results show that 80 per cent of newly introduced products fail to establish market presence after two years (Barrett, 1996). Indeed, cases involving international high profile companies are frequently cited to warn of the dangers of failing to utilise market research (e.g. Unilever's Persil Power and R. J. Reynold's Smokeless Cigarette).

Given the inherent risk and complexity, managers have asked for many years whether this could be reduced by market research. Not surprisingly, the marketing literature takes a market driven view, which has extensive market research as its key driver (Booz, Allen and Hamilton, 1982). The benefits of this approach to the new product development process have been widely articulated and are commonly understood (Kotler, 1998). Partly because of its simplicity this view now dominates management thinking beyond the marketing department. Advocates of market research argue that such activities ensure that companies are consumer-orientated. In practice, this means that new products are more successful if they are designed to satisfy a perceived need than if they are designed simply to take advantage of a new technology (Ortt and Schoormans, 1993). The approach taken by many companies with regard to market research is that if sufficient research is undertaken the chances of failure are reduced (Barrett, 1996). Indeed, the danger that many companies wish to avoid is the development of products without any consideration of the market. Moreover, when a product has been carried through the early stages of development it is sometimes painful to raise questions about it once money has been spent. The problem then spirals out of control, taking the company with it.

The issue of market research in the development of new products is controversial. The debate will continue for the foreseeable future about whether product innovations are caused by technology push or market pull factors. The issue is most evident with discontinuous product innovations, where no market exists. First, if potential customers are unable adequately to understand the product, then market research can only provide negative answers (Brown, 1991). Second, consumers frequently have difficulty articulating their needs. Hamel and Prahald (1994) argue that customers lack foresight; they refer to Akio Morita, Sony's influential leader:

> Our plan is to lead the public with new products rather than ask them what kind of products they want. The public does not know what is possible, but we do.

This leads many scientists and technologists to view marketing departments with scepticism as they have seen their exciting new technology frequently rejected due to market research findings produced by their marketing department. Market research specialists would argue that such problems could be overcome with the use of 'benefits research'. The problem here is that the benefits may not be clearly understood, or even perceived as a benefit by respondents. King (1985) sums up the research dilemma neatly:

> Consumer research can tell you what people did and thought at one point in time: it can't tell you directly what they might do in a new set of circumstances.

This is particularly the case if the circumstances relate to an entirely new product that is unknown to the respondent. New information is always interpreted in light of one's prior knowledge and experience. Rogers' (1998) studies on the diffusion of innovations as a social process argue that it requires time for societies to learn and experiment with new products. This raises the problem of how to deal with consumers with limited prior knowledge and how to conduct market research on a totally new product or a major product innovation. In their research analysing successful cases of discontinuous product innovations, Lynn *et al.* (1997) argue that firms adopt a process of probing and learning. Valuable experience is gained with every step taken and modifications are made to the product and the approach to the market based on that learning. This is not trial and error but careful experimental design and exploration of the market often using the heritage of the organisation. This type of new product development is very different from traditional techniques and methods described in marketing texts.

Knowing what the customer thinks is still very important, especially when it comes to product modifications or additional attributes. There is, however, a distinction between additional features and the core product benefit or technology, hence its emphasis in the framework (see Figure 7.1 on p. 120). The framework also places emphasis on the buyer rather than the consumer, for the buyer may be the end user but equally may not, as is the case with industrial markets. Moreover, for a product to be successful it has to be accepted by a variety of actors such as fellow channel members. The buyer is nonetheless a key actor. In industrial markets the level of information symmetry about the core technology is usually very high indeed (hence the limited use of market research), but in consumer markets this is not always the case. For example, industrial markets are characterised by:

- relatively few (information rich) buyers;
- products are often customised and can involve protracted negotiations regarding specifications;
- and, most importantly, the buyers are usually expert in the technology of the new product (i.e. high information symmetry about the core technology).

The above discussions imply the first proposition [1]: *that low information symmetry between buyer and consumer will limit the usefulness of any market research undertaken.*

In situations of low information symmetry consumers have difficulty in understanding the core product and are unable to articulate their needs and any additional benefits sought. Conversely, in situations of high information symmetry consumers are readily able to understand the core product and hence are able to articulate their needs and a wide range of additional benefits sought.

In the case of discontinuous product innovations, the use and validity of market research methods is questionable (von Hippel and Thomke, 1999; Elliot and Roach, 1991). As far back as the early 1970s Tauber (1974) argued that such approaches discourage the development of major innovations. It may be argued that less, rather than more, market research is required if major product innovations are required. Such an approach is characterised by the so-called technology push model of innovation. Products that emerge from a technology push approach are generated with little consideration of the market. Indeed, a market may not yet exist as with the case of the PC and many other completely new products. Frequently, consumers are unable to understand the technology in question and view new products as a threat to their existing way of operating. Martin (1995) argues that: 'customers can be extremely unimaginative. . . trying to get people to change the way they do things is the biggest obstacle facing many companies'.

Indeed, Ortt and Schoormans (1993) posit that potential consumers are not able to relate the physical aspects of a major innovative product with the consequences of owning and using it. Similarly, Hamel and Prahalad (1994) argue that while market research can help fine tune product concepts it seldom is the spur for an entirely new product concept. Consequently most conventional market research techniques deliver invalid results.

There are some well-known cases where companies have ignored the views of their customers. Chrysler, for instance, developed the original mini-van in the USA despite research showing that people recoiled at such an odd-looking vehicle. At the time of launch the product was revolutionary in appearance. Today, however, that basic design is the industry standard. The difficulty faced by product and brand managers is deciding whether feedback is surmountable scepticism or genuine lack of market acceptance. In technology intensive industries such as IT products consumers are often hesitant over any new purchase because of the rapid introduction of replacement and improved versions. This has led to adverse consumer reaction to rapid product improvements (Dhebar, 1996).

Discontinuous product innovations or radical product innovations frequently have to overcome the currently installed technology base – usually through displacement. This is known as the installed base effect. The installed base effect is the massive inertial effect of an existing technology or product that tends to preclude or severely slow the adoption of a superseding technology or product. This creates an artificial adoption barrier that can become insurmountable for some socially efficient and advantageous innovations. An example of this is the DVORAK keyboard, which has been shown to provide up to 40 per cent faster typing speeds. Yet, the QWERTY keyboard remains the preference for most users because of its installed base, i.e. the widespread availability of machines that have the QWERTY keyboard (Herbig *et al.*, 1995).

The idea of being shackled with an obsolete technology leads to the notion of switching costs. Switching is the one time cost to the buyer who converts to the new product. Porter (1985) notes that switching costs may be a significant impediment to the adoption of a new consumer product. It is the extent of the installed base that influences the cost of switching. This yields the second proposition [2]: *that a high installed base effect will limit the effectiveness of any market research.*

Using market research in the development of discontinuous new products: a conceptual framework

Firms can enable consumers to gain direct knowledge of their new products through first-hand experience. There are a variety of techniques used to trial new products such as using in-store samples, mailing samples to consumers' homes, product demonstrations and service samples and demonstrations.

In addition, consumers can gain information second-hand via company advertisements, packaging, point-of-purchase displays and Internet sites (Kardes,
1998). For innovative new products, one approach used is to gather together
innovators and early adopters to learn their views on a product idea. The problem here of course is that the focus of the study is on the characteristics of the
population rather than on the innovative product. For example, innovators
and early adopters are characterised by a high disposable income.

The examples cited in the previous sections serve to illustrate the
dilemma facing firms: market research may reveal genuine limitations with
the new product but it may also produce negative feedback on a truly
innovative product that may create a completely new market. This is why
there have been so many research studies in this area (Myers and Marquis,
1969; Rothwell, 1975, 1992; Cooper and Kleinschmidt, 1993; Hart, 1993).
And why there is such a wide variety of interesting cases of both successful
and less successful research projects, which are extremely valuable starting
points for analysis. Drawing on this literature an attempt has been made
to explore *under what circumstances does market research hinder the development of
innovative new products?* This has yielded a framework that provides useful
insights into both policy and operational issues. The impetus for the development of this framework was the realisation that substantial uncertainty
exists about decision-making based upon market research findings. This
uncertainty could be usefully divided into two independently identifiable
dimensions (see Figure 7.1) that can be described as:

1   Information symmetry about the core technology between producer
    and buyer.[1]
2   The installed base effect.

Quadrant 1: learning from lead users

Under these conditions market research teams are usually taught to collect
information from users at the centre of their target market. They conduct
focus groups and analyse sales data, reports from the field, customer
complaints and requests, and so on. Then they rely on their own creative
powers to brainstorm their way to new ideas. Teams that follow this
method assume that the role of users is to provide information about
what they need, and that the job of in-house developers is to use that
information to create new product ideas. The lead user process takes a
fundamentally different approach. It was designed to collect information
about both needs and solutions from the leading edges of a company's
target market and from markets that face similar problems in a more
extreme form. Development teams assume that knowledgeable users
outside the company have already generated innovations; their job is to
track down especially promising lead users and adapt their ideas to the
business's needs (von Hippel and Thomke, 1999).

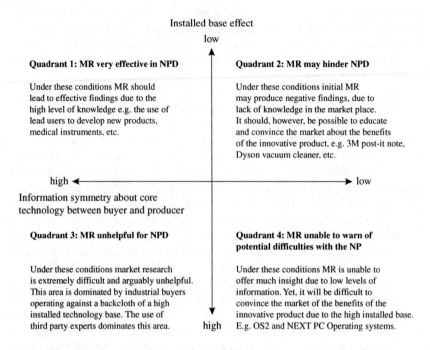

Installed base effect

low

Quadrant 1: MR very effective in NPD

Under these conditions MR should
lead to effective findings due to the
high level of knowledge e.g. the use of
lead users to develop new products,
medical instruments, etc.

Quadrant 2: MR may hinder NPD

Under these conditions initial MR
may produce negative findings, due to
lack of knowledge in the market place.
It should, however, be possible to educate
and convince the market about the benefits
of the innovative product, e.g. 3M post-it note,
Dyson vacuum cleaner, etc.

high ◄—————————————————————————► low

Information symmetry about core
technology between buyer and producer

Quadrant 3: MR unhelpful for NPD

Under these conditions market research
is extremely difficult and arguably unhelpful.
This area is dominated by industrial buyers
operating against a backcloth of a high
installed technology base. The use of
third party experts dominates this area.

**Quadrant 4: MR unable to warn of
potential difficulties with the NP**

Under these conditions MR is unable to
offer much insight due to low levels of
information. Yet, it will be difficult to
convince the market of the benefits of the
innovative product due to the high installed base.
E.g. OS2 and NEXT PC Operating systems.

high

Figure 7.1 Circumstances under which market research may hinder the
development of innovative new products

*Quadrant 2: Market research may hinder the development of innovative new
products*

Under these conditions it should be possible to educate and convince
buyers of the benefits of the innovative new product. Indeed, the key issue
here is how the benefits can best be articulated to the consumer. With a
low level of information symmetry, however, consumers will have difficulty articulating their needs.

It is clear that care has to be exercised when conducting market research
with innovative new products. In forecasting buyers' reactions to the innovative product, it is important to consider the product's or the technology's
functional performance. That is, what it will do for the end customer. Ortt
and Schoormans (1993) suggest that one approach is to use 'quasi-experts'.
These are consumers who have specific knowledge of related products
and of the usage context. In such cases it can be argued that there exists
improved information symmetry between the producer and consumer
vis-à-vis the core technology of the product.

Quadrant 3: Market research unhelpful

Under these conditions market research is extremely difficult and arguably unhelpful. This area is dominated by industrial buyers operating against a backcloth of a high installed technology base. In such cases it seems the role of third party intermediaries is significant. For example, within the information technology industry the role of IT analysts has been shown to influence notably the success or not of a new product launch (Vestey, 2000). The use of third party experts to analyse, comment and pass on their views to potential buyers dominates this area.

Quadrant 4

Under these conditions due to the high installed base effect it is extremely difficult to convince buyers to purchase the new product. Frequently the changes involved are substantial. This may be due to distribution arrangements, pricing, and internal reorganisation. This will be further compounded if the buyer has a low understanding of the technology. This may prevent the buyer from comprehending the potential benefits. Also consumers will have difficulty articulating their needs. As with Quadrant 2 the use of 'quasi-experts' may be helpful.

Discussion and management implications

This chapter has sought to investigate the extent to which, and under what conditions, market research may hinder the development of discontinuous new products. This is without question a controversial subject and has been debated for many years. There is ample evidence to suggest that market research can help firms develop improved products. However, there is also some evidence from the literature that firms may be concentrating their efforts on short-term activities, such as altering product attributes, rather than investing in more risky discontinuous product ideas (Morone, 1993; Lynn *et al.*, 1997). The conceptual framework presented here attempts to offer the practising product manager some guidance on when market research may be helpful and when its findings need to be treated with care. The operationalisation and subsequent empirical testing clearly are indicated in terms of further research. There are however some policy and operational considerations.

Policy considerations

If sufficient care is not exercised by managers market research can be used to support conservative product development decision-making. This chapter has highlighted the difficulty faced by many managers in the field of new product development. In many crucial new product development decisions, the course of action that is most desirable over the long run is not the best course of action in the short term. This is the dilemma addressed in the debate about short-termism. That is, an emphasis on cutting costs and improving efficiencies in the immediate future, rather than on creativity and the development of innovative new product ideas for the long term. What is of concern is not the desire to cut costs but the apparent disregard of the implications and damage that such policies may bring about, and in particular the neglect of the company's ability to create new business opportunities for the future well-being of the company (Trott, 1998).

Management needs to create new business opportunities for the company to ensure it is successful in the long term. Many of the world's largest chemical companies such as Shell, Du-Pont and ICI have for many decades afforded their scientists a percentage of their time to work on projects they considered interesting and of future potential value. This implies a willingness to invest in risky innovative new product ideas. The short-term view places emphasis on conservative decision-making supported by market research findings, with its in-built failings and inability to score innovative new products.

In neglecting attention to their creative activities companies may seriously affect their ability to innovate. Clearly there are many companies like '3M', 'Rubbermaid' and 'Lucas' (Robinson, 1997) that have not lost sight of the importance of creating new products for their future markets. Yet senior management thinking is often driven by management accountancy principles and the goal of ensuring costs are lower than revenues. The chief executive of BP Chemicals acknowledged that 'cost reduction is a miserable management job but conceptually it is easy' (Houlder, 1994). All of this leads to at best a management style that is reluctant to take risks and at worst a blame seeking culture that forces managers to ensure they can defend any decision they make; hence, the value given to market research findings.

Highlighting the importance of capabilities in R&D, technology development and channel management in new product development is not new, but Moorman and Slotegraaf (1999) suggest that product development is most effective when firms exhibit *both* capabilities. Furthermore, they argue that not all capabilities are valuable as single assets and more emphasis should be placed on the value of complementary assets.

Operational considerations

Arguably the need for innovation and creativity in market research is no less than in any other area of management. If market research is to provide useful effective information for helping organisations compete it needs to be capable of bringing forth insights and viewpoints from respondents. This is the challenge for market research. Some of the more recent research on new product development argues that successful firms are able to deploy complementary capabilities in technology and marketing (Moorman and Slotegraaf, 1999). This is based on the view that product development integrates inside out (R&D) and outside-in (effective management of customer and channel relationships) capabilities. It is worthy of note that capability rich firms are more apt to draw correct inferences about the value of external information (Cohen and Levinthal, 1990, 1994).

Market research can provide a valuable contribution to the development of innovative products. The difficulties lie in the selection and implementation of research methods. It may be that market research has become a victim of its own success, that is, business and product managers now expect it to provide solutions to all difficult product management decisions. Practitioners need to view market research as a collection of techniques that can help to inform the decision process. The conceptual framework outlined in this chapter should help product and brand managers to consider when and under what circumstances market research is most effective. The right sort of market research can be invaluable. The problem is that within consumer markets there are technology intensive and technology vacant industries. In many of the technology intensive industries such as telecommunications, computer hardware and software, these firms are able to utilise their industrial market heritage to balance the need for technology and listen to the needs of consumers. In technology vacant consumer markets, such as food and personal care, the danger is that the technology agenda is completely dominated by market research findings. Minor product modifications may keep a product and brand competitive in the short term, but if long-term growth is sought then more free-thinking and creativity needs to be afforded to the R&D department.

Conclusions

This chapter addresses two of the key influences on innovation, that being the market and the need for creativity (controlled chaos). The market is where firms compete; they must consider their customers and potential customers if they wish to be successful. Similarly, creativity is fundamental to innovation and its hindrance should be a cause for concern and needs to be addressed. This chapter has attempted to show that too much emphasis on the market and traditional interpretation of market possibilities can hinder the development of discontinuous new products. The extent to which market research is justified and whether companies should sometimes ignore their customers are key questions for firms. The conceptual framework offered should help companies to decide when market research findings may be helpful and when they may hinder the development of discontinuous new products.

Some of the most recent attempts to understand innovation theory have focused on the so-called *knowledge-based economy*, where intellectual capital has replaced production capital as the dominant force. This chapter has attempted to highlight the importance of technology and creativity in the innovation framework.

Notes

1   Christensen (1997) *The Innovator's Dilemma: When New Technologies Cause Great Firms to Fail*, HBS Press, Harvard MA, was awarded the *Financial Times* business book of the year award in 1999.
2   Discontinuous innovations often launch a new generation of technology; whereas continuous product innovations involve improving existing technology.
3   The technology push approach to NPD centres on trying to deliver the most effective technology available.
4   The term buyer is used here to take account of industrial markets, professional users and non-professional users (Pannenborg, 1986).

Part III

Internal innovation activities within the firm

The internal activities in the firm are the core of its innovative behaviour. This is the place of origin of the general ideas that develop into a strategy by meeting the external world and market conditions. The ideas for innovations, which the market or external actors may present are interpreted by the employees (including the managers). Many ideas come from the employees or managers. The process of developing and implementing the innovations also takes place in the organisation. This includes knowledge procurement, adaptations in the process and creative solutions of problems. These are important issues for understanding the internal process of what are the relationship between management and employees, the interaction processes and routines versus norm-breaking behaviour. Reflexivity occurs when problems in the innovation process lead to changes in routines. The internal activities are not only rational problem solutions, but also the inability to solve the problems, conflicts and manifestation of different partners' interests. In this section we present four chapters that analyse the internal innovation activities. One of the important issues that we start to discuss here is whether there is a fixed system for these activities or are they taking place in loosely coupled structures?

John Bessant starts by discussing this issue at a general theoretical level. His theme is whether routines for innovation activities are developed within the firm. The routine concept is discussed in relation to the notion of strategic reflexivity. Routines are the mechanism that enable interactions and negotiations, which are parts of the strategic reflexive innovation process. He describes how routines can develop, and discusses whether the innovation process can be managed. Innovations often fail for three reasons: (1) the lack of motivation to change; (2) incomplete models of the process to be managed; (3) poor execution of the process itself. The challenge to innovation management is to overcome these three problems. Bessant presents a model for how firms can learn to improve innovation management. This is seen as a learning based establishment of new routines for new challenges.

Anders Gustafsson, Bo Edvardsson and Bodil Sandén treat the internal innovation activities in relation to service innovations by analysing the case of the development of the travel experience (seen as a service product) in Scandinavian Air Services (SAS). The company has developed new, and redesigned old, services through allowing the customers to define the service process as opposed to defining it from the company perspective. Thus the innovation process is connected to the most important actor in the firm's strategic situation: the customer. The reflexive element of innovation is related to the customers' perception of the services delivered and which new aspects of these services – or completely new services – the customer wants. The innovation process at SAS is structured into four phases and it is described how the company's innovation activities are carried out. The authors finally discuss what can be generally learned from SAS's experiences.

Lars Fuglsang analyses innovation processes in a public organisation, a municipal home-help service. His object is an organisational innovation, a formation of new groups in the home-help service and the introduction of a new team manager profile. Through a qualitative case methodology he can demonstrate what has happened in the innovation process. He analyses the interaction pattern and different roles in the innovation process as well as the organisational system that was the result of the innovation process. He identifies a crucial role for the social entrepreneur, who is a team manager that breaks down the old, fossilised social/professional relations and facilitates the creation of new ones. The social entrepreneur introduces strategic reflexivity by helping the employees to identify the tasks towards the clients that the organisation has set up in its development goals and their role in the development of the organisation.

Marius Meeeus and Leon Oerlemans emphasise the implementation process and the interactions that take place there. They define organisational stakeholders and analyse their role in the innovation process. Theories of innovative organisation should emphasise the difficulties that innovation management faces due to the stakeholders' behaviour in the implementation of that process. The control over the process can not be hierarchical, but must be a two-way interaction. The authors establish three hypotheses, which say that implementation effectiveness depends on user participation and a positive attitude from the organisational stakeholders, who also should have low dissent on intervention outcome. The hypotheses are tested on automation implementation in six different organisations and were supported – except that in some cases disagreement between workers and management did not reduce the implementation effectiveness.

Jan Mattsson also deals with the implementation of innovations. He has studied how the innovation process has changed from planning and hierarchy to strategic reflexivity in social interaction. He has used a qualitative, interpretative method to analyses a process of implementing a PC based electronic control system for automatisation and certification of test facilities for marine engines. Quality and novelty in automation process are crucial. The case implied a struggle between two departments and he can thus analyse how an innovation process can be a conflicting interaction process. The employees and managers have certain roles in the innovation process and these roles may be contradictory. Thus, the innovation process does not always follow a hierarchical, rational consensus pattern, but can be a political process within the organisation.

8 Developing routines for innovation management within the firm

John Bessant

Introduction

It is a truism to say that innovation matters in the turbulent environments which characterise the current global economy. Renewing what a firm offers and the way it creates and delivers that offering (product and process innovation) is becoming an essential and core process necessary to the survival of the business (Brown *et al.* 2000). The alternative view – which sees innovation as an occasional 'luxury' option which firms can choose to indulge in or not – is not really tenable. Recent history points to a host of examples which highlight the risks involved in the latter course; even the largest firms are not immune to the dangers of complacency and lack of change. Analysis of the Fortune 500 firms over any extended period reveals how fragile the survival rates of modern businesses are, and the mortality rates for smaller enterprises are much worse. Put simply, the challenge today is one of innovation – continuous rather than occasional, and not just change for its own sake, but linked to a clear strategic focus.

The problem is that firms are not well adapted to the underlying core process – that of learning. Much current thinking on innovation sees it as the accumulation and deployment of knowledge – competencies – in strategic fashion (Hayes *et al.* 1988; Cohen and Levinthal 1990; Pisano 1994; Figuereido 2000). But – despite an extensive prescriptive literature on 'the learning organisation' – most firms do not learn very well. They are typically characterised by various mechanisms which aim to preserve stability, to damp out signals for change and to inhibit the development of novel responses to both new and recurring challenges in their environment (Garvin 1993). Even those which do well in terms of learning may find themselves caught out; learning is a dynamic process and the parameters within which it takes place need themselves to be reviewed and adjusted on a regular basis. Christensen's studies of the hard disk drive industry indicate that under certain conditions following good practice – for example, staying close to key customers and developing innovations in response to signals emerging from this relationship – may actually be the wrong thing to do. His view of 'disruptive technologies' suggests that at times firms need to respond to weak signals which apparently go against the whole basis of an existing business operation – in other words, they have to reset their learning parameters (Christensen 1997).

All of this places considerable emphasis on the question of managing innovation, and on the development of learning behaviours – routines – within the enterprise which enable it to take place. This chapter explores this issue and looks at some key routines associated with successful innovation at firm level.

Strategic reflexivity and the emergence of routines

A key theme throughout this book has been that of strategic reflexivity and this is a highly relevant concept for the discussion of routines. In essence the process of management involves one of interpretation of market and technical opportunities and the development of strategic responses arising from this interpretation. Structures and procedures emerge not as an accident but as the consequence of negotiated and rehearsed activities within which the roles and behaviours necessary to enable innovation are worked out.

The way in which routines emerge can be seen as one which derives from fundamental values and beliefs – 'taken for granted' assumptions about 'the way we do things around here'. These shape behaviours which in turn become shared norms – and lead to the construction of artefacts – structures, systems, procedures, symbols – which reinforce the underlying beliefs into the organisational culture (Schein 1985). In response to external stimuli – the threats and opportunities which constitute the triggers for innovation – this culture needs to change and adapt – and the process of doing so involves articulation and exploration of new options followed by extensive rehearsal and refinement.

The process is often one of trial and error but can be influenced and enabled by management action – particularly in the use of incentives and through the medium of new working approaches – tools and techniques – which offer formalised ways of employing new behaviour patterns (Nelson and Winter 1982).

Levitt and March describe routines as involving established sequences of actions for undertaking tasks enshrined in a mixture of technologies, formal procedures or strategies, and informal conventions or habits. Importantly, routines are seen as evolving in the light of experience that works – they become the mechanisms that 'transmit the lessons of history' (Levitt and March 1988). In this sense, routines have an existence independent of particular personnel – new members of the organisation learn them on arrival, and most routines survive the departure of individual routines. Equally, they are constantly being adapted and interpreted such that formal policy may not always reflect the current nature of the routine – as Augsdorfer points out in the case of 3M (Augsdorfer 1996)

An important aspect of routines is that they are not necessarily repetitive; rather, their execution does not require detailed conscious thought.

The analogy can be made with driving a car; it is possible to drive along a stretch of motorway whilst simultaneously talking to someone else, eating or drinking, listening to, and concentrating on something on the radio or planning what to say at the forthcoming meeting. But driving is not a passive behaviour; it requires continuous assessment and adaptation of responses in the light of other traffic behaviour, road conditions, weather, and a host of different and unplanned factors. We can say that driving represents a behavioural routine in that it has been learned to the point of being largely automatic.

In the same way, an organisational routine might exist around how projects are managed, or new products researched. For example, project management involves a complex set of activities such as planning, team selection, monitoring and execution of tasks, re-planning, coping with unexpected crises, and so on. All of these have to be integrated – and they offer plenty of opportunities for making mistakes. Project management is widely recognised as an organisational skill, which experienced firms have developed to a high degree but which beginners can make a mess of. Firms with good project management routines are able to codify and pass them on to others via procedures and systems. Most important, the principles are also transmitted into 'the way we run projects around here' by existing members passing on the underlying beliefs about project management behaviour to new recruits.

For our purposes the important thing to note is that routines are what makes one organisation different from another in how they carry out the same basic activity. Each enterprise learns its own particular 'way we do things around here' in answer to the same generic questions – how it manages quality, how it manages people, etc. 'How we manage innovation around here' is one set of routines which describes and differentiates the responses which organisations make to the challenge of change.

Can we manage innovation?

It is reasonable to begin by asking whether or not innovation can really be managed. After all it is, by definition an uncertain and risky process whose outcomes cannot be predicted. And the evidence of failure, even amongst well-managed and resourced organisations, is worrying; both product and process innovation are high risk activities. It is tempting to take the view that with so many variables involved – technological development, market/demand variations, internal organisation and external environmental influences, etc. – it is simply impossible to design and execute changes in products or processes. When firms do succeed this is simply a matter of lucky accident.

Whilst there is considerable risk and uncertainty involved, it does appear that some firms are able to succeed on a consistent basis. Far from getting lucky once they seem able to repeat the trick on a regular basis – not with

100 per cent success, it is true, but with a favourable balance of success over failure. It is possible to attribute this to luck but, as Pasteur once said, 'chance favours the prepared mind'. It is more likely that these organisation succeed because they have learned particularly effective behavioural strategies for organising and managing innovation, and have embedded these in the operating structures and procedures of the firm (Van den Ven *et al.* 1989; Rothwell 1992).

Examples abound in the literature and suggest that there are lessons which can be learned from these firms about ways of managing which seem to favour the operation of the innovation process (Nayak and Ketteringham 1986; Grindley *et al.* 1989; Mitchell 1991; Gallagher and Austin 1997). Identifying, articulating and reinforcing these behaviours – and providing the structures and procedures with which they become embedded in the organisation – is the primary task of innovation management.

Three key challenges in innovation management

Innovation fails for many reasons but we can begin to identify certain types of failure which are repeated in different forms – and around which behavioural routines might be developed to help increase the chances of more successful outcomes. In the remainder of this chapter we will consider three examples of such clusters and relevant routines which are associated with improved innovation management. These are:

- The lack of motivation to change
- Incomplete models of the process to be managed
- Poor execution of the process itself

Lack of motivation to change

The first problem in managing innovation concerns picking up and responding to the signals for change in the organisation. Innovation is essentially about some combination of demand pull and knowledge push, but unless the organisation is able to detect stimuli from either of these directions there will be little to trigger the process of change. The problem may be one of lack of awareness – for example, many smaller firms are often insulated from a clear understanding of developments in wider markets. Insulation can also come from the effect of trade and other barriers which provide a measure of shelter and protection from the market forces operating elsewhere and which carry signals about the need for change. In other cases firms may simply be scanning in too limited an area – what Carter and Williams termed a 'parochial' orientation (Carter and Williams 1957). As Utterback reports, the risks here are often substantial since in cases where industry transformation through technological change takes place the trigger signals often arise from outside the established industry (Utterback 1994).

The problem is not simply one of picking up signals; it also concerns the way in which firms choose to respond to them. Much is made of the turbulence of the current business environment and the need for constant innovation, but this is to ignore that fact that most enterprises are organised around resisting change and promoting some form of equilibrium. Change is risky, consumes resources and threatens to upset established arrangements, so there is often a self-damping character to responding to signals. Firms develop routines for preserving the status quo just as efficiently as they do those for innovation.

One example is the famous 'not invented here' effect where firms react to signals about a potentially significant change by dismissing it because it does not fit their particular perspective. The case of the telephone is a good illustration of this; Alexander Bell first took his new idea to the major communications firm, Western Union to see if they would help him commercialise it. Their written reply, a few days later, suggested that 'after careful consideration of your invention, which is a very interesting novelty, we have come to the conclusion that it has no commercial possibilities . . . we see no future for an electrical toy.' Within four years of the invention there were 50,000 telephones in the USA and within twenty years there were 5,000,000. In the same time the company which Bell formed, American Telephone and Telegraph (ATT) over the next twenty years grew to become the largest corporation in the USA, with stock worth $1000/share. The original patent (number 174455) became the single most valuable patent in history.

This pattern of selective perception and filtering is something which Dorothy Leonard draws attention to in her discussion of how core competencies can switch to being core rigidities in large organisations (Leonard-Barton 1995). That is, the very things which help a company build a successful competitive position become those which act as filters on its perceptions of subsequent stimuli. So firms may not assess the threat or opportunity associated with a new technological or market opportunity because they frame the stimulus in terms of their existing pattern of core competencies.

Christensen makes a related point in discussing 'the innovator's dilemma'. He argues, based on his work in the hard disk drive industry, that under certain conditions the signals coming from the existing customer base may not be reflecting the new growth and innovation opportunities. He argues that firms should not always follow the classical business advice of staying close to their customers but on occasions should look for different fringe customers since it is this group which has the new stimuli which will define the new possibilities for innovation (Christensen 1997).

This brief discussion highlights the difficulties in entering into the innovation process; in order to manage past this barrier firms need to develop routines which actively search and challenge the status quo. Table 8.1 gives some examples.

Table 8.1 Routines for dealing with barriers to innovation motivation

Barriers	Enabling routines
Insulation or isolation	Active search behaviour Systematic comparison – benchmarking and other forms of comparative measurement Out of industry benchmarking Industrial visits/demonstration projects Forecasting Competitor analysis
Denial – it's not really happening	Evidence-based management Measurement – especially comparative data from benchmarking and market research Constructed crisis
Core rigidities/not invented here	Competence auditing and review

Limited understanding of the innovation process

The second area in which problems can occur lies in the ways in which the process of innovation is understood within the firm. As we saw earlier, and following Schein it is possible to argue that behaviour results from core beliefs and values, and that this behaviour creates artefacts – structures, procedures, symbols, etc. – which reinforce and embed particular behaviour patterns (Schein 1984). So it follows that the ways in which firms see the innovation process will have a marked influence on the ways in which they choose to manage it. With that view of only partially understanding the resulting innovation, management routines may be limited in their effectiveness. Table 8.2 gives some examples.

Problems in execution

The third area of difficulty in managing innovation lies with the actual execution of the process – and it is here that routines become of central importance. Clearly some routines are more effective than others and represent the learning which has taken place over an extended period of time to develop firm-specific ways of dealing with innovation challenges. In an era of benchmarking, understanding how different firms use different routines for the same common renewal process offers considerable potential for inter-firm learning.

We can usefully map these execution routines on to a simple model (Tidd *et al.* 1997) of the innovation process which sees it as a process with the following stages:

- scanning the environment for relevant signals and processing these
- selecting projects which have a good strategic fit
- monitoring and managing projects through the various stages of development

- deciding where and when to stop projects, and where and when to accelerate them
- reviewing and capturing learning from completed projects

For many enterprises this kind of activity is often carried out on an ad hoc basis, and not in systematic fashion. But each of these stages requires particular capabilities and these can be enabled by specific behavioural routines. Table 8.3 gives some indicative examples.

Table 8.2 Problems of partial understanding of the innovation process

If innovation is only seen as . . .	*. . . the result can be*
Inventive ideas	Idea which fails to meet user needs and may not be accepted
The province of specialists – 'designers' R&D laboratory, creative specialists, etc.	Lack of involvement of others, and a lack of key knowledge and experience input from other perspectives
Meeting customer needs	Lack of technical progression, leading to inability to gain competitive edge
Technology advances	Producing products which the market does not want or designing processes which do not meet the needs of the user and which are opposed
Only about 'breakthrough' changes	Neglect of the potential of incremental improvement
Only associated with key individuals	Failure to utilise the creativity of the remainder of employees, and to secure their inputs and perspectives to improve innovation
Only internally generated	The 'not invented here' effect, where good ideas from outside are resisted or rejected
Only externally generated	Innovation becomes simply a matter of filling a shopping list of needs from outside and there is little internal learning or development of technological competence

Source: Tidd *et al.* (1997) *Managing Innovation: Integrating Technological, Market and Organizational Change*, p. 31, Table 2.3 © John Wiley & Sons Limited. Reproduced with permission.

Learning to improve innovation management

Developing routines is, as we have said, a learning process which derives from the trial and error experiences of firms working at innovation (Pavitt 1990). Such learning does not take place automatically and may not even take place at all. Thus it is important to include in the suite of routines for good innovation management practice those which deal specifically with the challenge of learning and its management. Firms have a choice

Table 8.3 Key capabilities in innovation and routines to support their development

Stage in the process	Key capabilities required	Indicative routines
Scan	Scanning and searching Forecasting Experience-sharing and networking Signal processing Communicating and connecting	Market and technology research Forecasting activities involving multiple disciplines, internal and external people and range of tools and techniques – scenarios, Delphi, trends, foresight, etc. Active networking Benchmarking (Womack, Jones *et al.* 1991; Oliver 1996) Working with users (Von Hippel 1988; Herstatt and von Hippel 1992) 'Reverse engineering' and other competitor studies 'Voice of the customer' techniques Quality function deployment (Hauser and Clausing 1988; Shillito 1994)
Strategy	Signal processing Visioning Analysing Aligning Positioning (internally and externally) Selecting Planning	High involvement strategy process (Francis 1994) Matching resource driven and external positioning – 'inside out' and 'outside in' perspectives (Teece and Pisano 1994) Prototyping, concept testing and other feasibility studies Policy deployment
Acquisition	Make/buy analysis Articulating needs Search Evaluating options Absorption and assimilation	Development of a broad portfolio of technology sourcing routes – from internal R&D through to technology transfer Collaboration (Dodgson 1993) Networking (Schmitz 1998) Strategic alliances Joint ventures Technology transfer management (Levin 1993; UNIDO 1995)
Implementation		Development funnel – management by stage gates (Wheelwright and Clark 1999; Cooper 1994) Parallel market and technology development (Smith and Reinertsen 1991) Project team building (Thamhain and Wilemon 1987; Kharbanda and Stallworthy 1990; Katzenbach and Smith 1992; Holti *et al.* 1995) Project management structures and operating procedures (Wheelwright and Clark 1999) Cross-functional working Early involvement

as far as learning is concerned. They can – like too many examples – carry on, ostrich-like, ignoring the signals about the need to adapt, or they can undertake planned and systematic learning. In the former case there is considerable evidence to highlight the risks. Firms which don't recognise the need for change – especially those which find ways of reassuring themselves that their position is strong and they are insulated from the need to adapt – run the risk of being overtaken by events (Tushman and Anderson 1987; Utterback 1994).

Equally there are organisations which recognise the need and develop planned approaches to capture and manage learning. For example, in the case of computer aided production management systems in one multi-divisional firm a 'pioneer' team was set up. This group had the task not only of managing the first implementation project, but then capturing the important lessons learned and carrying those forward to the next implementation – and so on. By the fourth or fifth the process had become smooth and successful, and the company had absorbed valuable lessons about both the technology itself and how to manage its implementation (Leonard-Barton 1992).

At the outer limits we might expect to find what are sometimes called 'learning organisations' – firms which are structured and operate in such a way as to make change a way of life with a high degree of involvement of everyone in the process of continuous learning (Pedler *et al.* 1991; Garvin 1993; Leonard-Barton 1995).

One way of representing the learning process which can take place in organisations is to use a simple model of a learning cycle – see Figure 8.1. Here learning is seen as following a pattern of reflecting on experience, building up conceptual models and then testing out their validity by experiment. The cycle can be entered at any point; the important message is that learning only takes place when the cycle is completed. Many organisations are strong in terms of the experimentation and resulting experience – what they often lack is the time and space in which to reflect and the conceptual models and frameworks with which to make sense of the things which they are trying out.

If we are to learn well from innovation management then we will need routines to support:

* *structured and challenging reflection on the process*: what happened, what worked well, what went wrong, etc.
* *conceptualising*: capturing and codifying the lessons learned into frameworks and eventually procedures to build on lessons learned
* *experimentation*: the willingness to try and manage things differently next time, to see if the lessons learned are valid
* *honest capture of experience:* (even if this has been a costly failure) so we have raw material on which to reflect

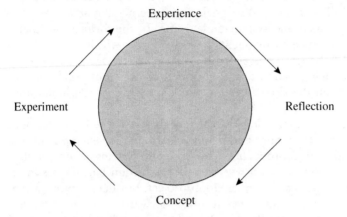

Figure 8.1 Simple model of a learning cycle (after Kolb)

Source: Kolb and Fry 1975

Effective learning from and about innovation management depends on establishing a learning cycle around these themes. To help with the process there is a variety of tools and mechanisms, and Table 8.4 gives an overview of some of these.

Developing new routines for new challenges

The routines described in the previous section represent an emergent model or reference framework for 'good practice'. They are the product of a learning process which defines effective responses to the challenges of managing innovation within the current 'envelope' of the firm's experience. But we need to recognise that innovation, by its very nature, is a dynamic process, and the demand side challenges and supply side opportunities will change on a continuing basis. It is about a constantly mutating puzzle in which the supply (technology push) and demand (need pull) side factors are always changing. New configurations – for example, the emergence of a technological field like the Internet with its widespread application potential – pose new managerial challenges and require the development of some routines, the acquisition of some new ones and the letting go of others which are no longer appropriate. Table 8.5 indicates some emerging challenges around which firms are likely to need to develop new routines or significantly adapt existing ones.

Learning to learn

Lastly – and of particular significance – is the point that effective innovation management needs to include a set of routines which deal with the

Table 8.4 Routines for enabling learning within the innovation process

Reflection	Conceptualise	Experiment	Experience
Post-project reviews	Theories and models	Pilot projects	Capture experience – on video, via diaries, project
Benchmarking	New structures and process designs	Testing and prototyping	records, photographs, etc.
Structured audits	Formal planning reviews	R&D activities	Sharing experience – via display, direct exchange, etc.
Project evaluation	Training and development	Designed experiments and simulations	Documentation and display
Measurement			Measurement

development of learning capacity itself – the ability to define, articulate and embed new routines and to delete those from the repertoire which have no further relevance in the new environment. In terms of learning models this corresponds to 'double loop' or 'generative' learning and implies the capacity to step back and reframe the innovation problem and the ways in which its management might be approached (Argyris and Schön 1970; Senge 1990).

This is the kind of learning behaviour which often characterises firms in a state of crisis when they have to make fundamental shifts in their thinking and reset the entire organisational culture. There are many stories of dramatic turnarounds resulting from such traumatic events but these are the lucky survivors; for many firms waiting until a crisis of this magnitude forces a major rethink may be too late. For this reason there is a growing body of experience suggesting the value of routines which explore and rehearse alternative approaches – for example, the extensive use of scenario planning within Shell or the deliberate construction of crisis thinking at Xerox and in the Korean shipbuilding industry (Camp 1989; de Geus 1996).

Conclusions

Although a highly uncertain and complex process, innovation is capable of management. This chapter has looked briefly at the concept of behavioural routines as the core mechanism through which such management of the process takes place within organisations. Such behaviours represent firm-specific responses to the challenge but as they become codified and institutionalised into routines, so they become visible to others and offer the possibility of learning and experience sharing.

Table 8.5 Emerging challenges in innovation management

Emerging issue	Implications for innovation management routines
How to build temporary and flexible structures for innovation	Increasing teamworking with professionally skilled and mobile participants (Francis 1997) Managing devolved autonomy within cells, 'fractals' and other sub-units which increasingly substitute for formal structures (Warnecke 1992)
How to build high involvement innovation	Continuing emphasis on training and skill development Development of learning skills Policy deployment (Bessant and Francis 1999) Enabling structures for continuous improvement (Imai 1997) (Bessant and Caffyn 2001)
How to explore innovation space more effectively – new combinations of products, processes and positioning	Build cross-functional links between different groupings Create challenge culture – internal and external review teams, panels, etc.
How to manage innovation across networks, chains and systems	Shared learning systems (Rush *et al.* 1997) Learning networks (Bessant and Tsekouras 2001) Supply chain development (Hines 1994) Trust management (Humphrey and Schmitz 1996)

Routines and strategic reflexivity

Recent years have seen an upsurge in critical interest in the models used to describe the operation of the innovation process, and in particular there has been a growing recognition of the interactive nature of the process. From simple linear models of push and pull we have moved to variations on an interactive theme; this is well charted by Rothwell in his review of five generations of innovation thinking (Rothwell and Bessant 1992). But it is possible to imagine that a sixth model may be emerging which sees the challenge extending beyond the interaction of technology and market forces, aided and sustained by powerful networking technologies. This 'sixth wave' is characterised by the addition of the dimension of social interaction both inside and outside the enterprise. Increasingly – as this book has argued – the process of innovation is likely to be one involving a high degree of negotiation and construction by those involved and affected. We have seen this in the context of internal changes – such as the introduction of IT systems which require a high degree of consultation and user-involvement in design. Outside the firm it has been clear for some time that the interplay of social forces with market signals is critical; innovation here too is the result of a negotiated process.

This places considerable emphasis on the development of organisational behaviours – routines – which are capable of enabling such negotiation and interaction. In the introduction the point was made that 'strategic reflexivity is a condition where people can do something about problems through innovation'. Embedded and successful routines provide the channels for making this happen – and their articulation and reinforcement represents a key challenge for organisational development.

9 Mapping customer behaviour

A key to successful new service development and innovation

Anders Gustafsson, Bo Edvardsson and Bodil Sandén

Abstract

The process of developing and launching new services is vital to the competitiveness of service organisations. However, service innovation and new service development are poorly researched and understood areas within services marketing and management. Most of the prior theoretical research and empirical studies has been characterised by the adoption of frameworks and models from manufacturing. Yet, these models do not consider important aspects of the service logic such as its intangibility, customer heterogeneity, customer co-production, and impossibility to keep in stock. The process nature of services and the role of customers as actors and part-time employees in value-creating processes are the core of what we in this chapter call the service logic.

The airline industry is going through dramatic changes. This is due foremost to altered customer demands and expectations but also to deregulation of the airline business and the introduction of information technology in new ways. Many airline companies have lost track of the true needs of their passengers and are trapped in outdated views of what airline services are all about. This article illustrates how Scandinavian Airlines System (SAS) has carried out thorough investigations into the concerns of the customers throughout the entire travel experience. Based on the results, SAS has developed new and re-designed old services. One of the key issues in the article is the way SAS has chosen to develop the new services, namely to allow their customers to define the process as opposed to defining it from the company perspective. This was done by observing customer behaviour with the help of video-recording techniques. The approach is applicable in many service companies in the development of new services.

Introduction

The term innovation is not a clearly defined or used concept. Rogers (1995, pp. 1–20) sees innovation as a process that starts with the invention of a new element. The invention leads to the idea of practical development of the innovation into commercial use. The innovation must be both practically usable and profitable. Others will join and a general diffusion process will take place. Most often the term innovation is used to describe this overall process and its various parts. The innovation theory is based on Schumpeter's (1934) theory, which focuses on the renewal and growth in an economy. Sundbo (1998a) uses 'innovation' to describe 'the effort to develop an element that has already been invented, so that it has a practical-commercial use, and the attempt to gain the acceptance of this element' (p. 12). We agree with Sundbo who argues that a distinction between an innovation and a normal process of change is difficult to make. The concept of innovation will lose its relevance if it is reduced to merely denoting any social or economic change. Even incremental innovations must include a qualitative leap through the introduction of a new element. 'Thus, the diffusion of innovations denotes an economic process of development – as opposed to an economic growth process, which merely is a quantitative increase of the existing elements' (p. 14). Sundbo concludes that the general innovation concepts are applicable to services and that 'innovation has become increasingly important, both in the daily function of the service companies, and in the thinking of service company management' (p. 348).

The process of developing and launching new services is vital to the competitiveness of service organisations. However, service innovation and service developments are poorly researched and understood areas within services marketing and management (Grönroos 1990, Martin and Horne 1993, Edvardsson, *et al.* 1995). Most of the prior theoretical research and empirical studies have been characterised by the adoption of frameworks and models from manufacturing. Yet, these models do not consider important aspects of the service logic such as its intangibility, customer heterogeneity, customer co-production, and impossibility to keep in stock. Service processes involve some form of interaction with customers before the service is complete. The introduction of a new service must be evaluated in the light of altered customer perceptions and changes in the actions of the customers during the delivery process. New service development and innovation concepts, theories and models must be based on an understanding of the service logic.

This chapter contributes to the theoretical framework, described in Chapter 1 of this book by focusing on customer behavior and how new services and innovations may contribute to the customer's own value creation. The focus is on how a service organisation can design service offerings and service processes to support customers in their interactions and relationships with the service provider.

Strategic reflexivity is seen as a strategic response to change-processes on the market and in society. We argue that these changes, or at least key aspects of these changes, may be described and understood by observing our customers' behaviour and by interviewing them about what they want to do during a service process, what needs they have and how they want these to be met. One way of responding to external changes is to understand and respond to the changing needs, wants, and behaviour of existing customers. In most service industries, customers are becoming more demanding, and processes more and more transparent. Information provided via the Internet on different alternatives gives more power to the customers. This requires that service companies listen to and understand changes in the customers' value creation processes. Furthermore, companies must be able to respond to changing customer requirements by introducing new services and innovations on a continuous basis. This is a prerequisite for customer satisfaction, loyalty and long-term profitability. Innovations in the new economy are based on knowledge and core competencies, flexibility, service-value focus, customer perspective and a proactive response to changing market conditions. In this chapter, we will describe how SAS forms the basis for applying strategic reflexivity by mapping and understanding their customers' actual and preferred behaviour during parts of the travel process. Our case study clearly illustrates the key role of customer ideas and behaviour data in the innovation process. Furthermore, it illustrates the importance of the interactions between the customers and the employees during the innovation process as well as the customers' interaction with technology forming the service infrastructure. We can label this chapter: 'Innovation as strategic reflexivity in services – the customer active paradigm'.

Aim and structure of the chapter

The aim of this chapter is to contribute to a better understanding of the new service development and innovation process based on the service logic. We will emphasise the interaction between customers and employees and the customers' contributions to new services and innovations. Innovation is seen as an interaction process between customers and the service provider. We will illustrate how Scandinavian Airline Systems (SAS) has tried to build its development activities on what customers do and want to do when they travel. The company used thousands of hours of video recordings to closely examine passenger activities throughout the entire travel experience. By these means, key customer needs and concerns were identified and turned into a foundation for design principles that were to shape company development strategies and innovations. The following description is based on the previously mentioned service model, however with emphasis on the idea phase. This is due to the fact that principally all SAS activities mentioned in the case can be ascribed to that particular phase.

The structure of the chapter is as follows: we begin with a theoretical frame of reference in which we define the service concept and describe key service characteristics. We also give a brief overview of research on the new service development process and a summary of the knowledge of new service development and innovation. We then turn to our case study within SAS and start with a description of today's airline industry situation. A presentation of the project carried out within SAS follows this part including issues such as how and why the project was initiated and how the customer data was collected and analysed. Conclusions from the study are presented and discussed in relation to previous research, and the general utility of the findings is also considered.

Theoretical frame of reference

Defining services

Services could be viewed as a part of the wider 'product concept'. A product can consist of commodities, services and computer software or of a combination of these (Edvardsson 1997). Another term used is service product, which implies that it is primary intangible core attributes the customer buys (Johne and Storey 1998). Following Grönroos:

> A service is an activity or series of activities of more or less intangible nature that normally, but not necessarily, take place in interaction between the customer and service employees and/or physical resources or goods and/or systems of the service provider, which are provided as solutions to customer problems.
>
> (Grönroos 1990, p. 27)

The quote makes us aware of the fact that the service aims at solving a *perceived customer problem*, the so-called customer result. Another important part of the definition is that a service appears through a *process of activities*. In his definition Bateson (1989, p. 6) emphasises that a service is an experience and that the service is made up of a number of benefits: 'services deliver a bundle of benefits to the consumer through the experience that is created for the customer'. A service is the customer's experience of the process (Edvardsson 1997). Activities and resources that the customer does not notice or experience do not exist. In the end it is the customer's experience of resources, activities and processes that form the perception, for example with regard to quality and value. With this view it becomes important to understand what creates or makes the experience (Edvardsson 1996).

Based on the above stated definitions, a service can be characterised as a chain of (sequential and parallel) value creating activities or events, which form a process. In this process, the customer often takes part by performing different activities in interaction with the employees of the service company (other customers or equipment) for achieving a special result (Edvardsson 1996).

Substantial attempts have been made to distinguish the characteristics that differentiate services from traditional manufactured products. Often referred to is Zeithaml *et al.* (1985). They refined four factors including intangibility, co-production, heterogeneity and perishability. Another study which aimed to identify what characterises services and partly separates services from physical products, identified four characteristics (Edvardsson and Thomasson 1989, pp. 23–4):

- *Immateriality*: Services are usually more or less abstract and immaterial, which makes them difficult to inspect and examine before buying them. Services do not have a lifetime, they only exist under a short period and therefore, they usually cannot be stored or saved.
- *Co-production*: Services are usually partly produced, delivered, consumed and marketed at the same time. This has the effect that quality must be built in as the service is developed (Gummesson 1991). 'Quality is not engineered in the manufacturing plant, then delivered intact to the customer. In labour-intensive services ... quality occurs during service delivery' (Parasuraman *et al.* 1985, p. 42).
- *Customer as a co-producer*: Services usually involve the customer in the role of co-producer by way of: (1) supplying information and other inputs to the process, (2) performing one or several activities in the service process or (3) marketing the service through talking to others about impressions and perceptions. Therefore, the customer should be seen as a part time employee or a resource that contributes with, among other things, knowledge and information.
- *Heterogeneity*: Services are heterogeneous, e.g. depending on how capital, customer or personnel intensive the service is. Also, as the customer takes an active part and has his own unique needs, demands and values, it will affect both the process and the result. This makes it hard to standardise and control the process as well as the outcome.

These characteristics are known as the service logic. Such distinctions propose that special considerations must be taken into account in the design of new services and improvement of existing ones. This in turn indicates that the new service development process has special features that are to some extent different from a traditional product development process for manufactured goods.

The process of new service development

Four essential characteristics of an effective development process for new services have been identified including objectivity, precision, fact-driven and methodologically based procedures (Shostack 1984). These characteristics do not seem unattainable. However, service suppliers do not seem to use any sophisticated or formal development procedures.

Johne and Storey (1998) find it surprising that there have not been more attempts to design a specific service development model. The majority of the early models of the new service development processes are often strongly sequential in nature. According to our experiences this is not a correct depiction. We found that the phases often overlap. It is also by looking at the models, that we determine whether or not they are based on the service logic. A description of the differences between services and commodities is often provided in an introductory part but rarely could one detect these special characteristics in the models.

One of the more cited and detailed models of new service development, built up by 15 phases based on previous models, is provided by Scheuing and Johnson (1989). Edvardsson (1996) is quite sceptical about this model since the authors have not presented the results from an inquiry they did to test the model. Neither did they provide any further details about how the model was changed, based on the empirical survey.

An alternative service development process could be described in four phases. First, an analysis and valuation of customer benefit is undertaken to guarantee that the development work is guided by total quality for the customer. Second, an analysis of the competition and the features of a competitive augmented service offering, including the production and delivery process as well as market communication, is entered into. The third step is the development of the core package, supporting services/products and secondary services/products and its materialisation through the production and delivery process. The final step comprehends the planning of supportive market communication, which not only should inform the customers and convince them of trying the service but also build up a positive image that supports the service (Grönroos 1990).

People's involvement in new service development is crucial. There are three groups of individuals that need to be engaged in an effective development project: the development staff, the customer-contact staff and the customers.

One of the key concerns in new service development is the lack of skilled and experienced *development staff* (Drew 1995, Johne and Harborne 1985). There is a need to create a 'taskforce' that is separated from day-to-day functional pressures. It is important that neither marketing nor operations dominate new service development. The approach should be cross-functional. Edvardsson and Gustafsson (1999) broaden this view by stating that the approach should be multi-functional, which also includes, above the cross-functionality, experts from different areas.

There are a number of benefits that could be met in terms of employee involvement in new service development. First, the *customer-contact staff* can help to identify customer requirements and needs. Second, the front office's involvement increases the likelihood of positive implementation. Third, it helps stop process efficiency considerations overtaking the needs of customers. And finally, it can lead to employees treating customers better. Employees, however, are often unwilling to get involved in development activities due to their ordinary workload (Scheuing and Johnson 1989). Internal marketing is of vital importance in order to sell the idea to the customer.

The final group includes *the customers*. Their involvement is essential since it is their needs, demands and requirements that the service supplier wants to satisfy. It is not easy for the company alone to define these needs, nor is it always easy for the customers to articulate them. Customers also need to be involved in order to develop a process that is customer friendly and adapted to human logic. The best person to judge this is the customer himself. In general, the more involvement by customers the better, though on the whole customer involvement in service development has been found to be relatively low (Martin and Horne 1995).

In general, many service companies do not seem to have any trouble generating new service ideas (Easingwood 1986). However, there is an overall lack of radical innovation in services. According to Johnson (1998) only about 10 per cent of all new services on the market could be defined as innovations.

Due to the ease of copying services, competitors have always been identified as an important source of ideas for new services, even more important than customers are. This might explain the lack of innovation in the service industry. There is a danger in focusing mainly on 'me-too' services, which are reactive and defensive in nature. Edvardsson and Gustafsson (1999) stress the importance of using a multi-method approach, meaning that there is a need to use more than one market research method covering the same issue. The benefit is that we then receive multiple views of the customers' problems. Consequently, we have a better understanding of what service the customers want.

A popular type of organisation for service development is project (Terril 1992). These projects must be supported by managers (Mattsson 1992). To succeed it is vital to encourage internal communication and involve people from several different functions and with different expertise. There is a need to include customers in the process of developing a concept, business analysis, testing and launching. In a study of organisational aspects on new service development four key areas for success were found (Thwaites 1992). These include timing, a clear business concept and suited strategies, competent and engaged co-workers, and a working communication.

A number of success factors in order to create and secure quality in new service development have been identified (Edvardsson and Gustafsson 1999, p. 24). Among these, certain organisational factors are mentioned. According to the authors it is not enough to have a deep understanding of the customers' needs, expectations etc., one also has to *create a service culture* within the organisation. New development projects should be carried out in *multifunctional teams* in *close cooperation with real and demanding customers on their premises*. The authors also argue the need for *communication*, internal as well as external. This is about getting the message through to as many employees as possible and at the same time getting as much input into the process as possible from customers. This is one of the key factors. Another one concerns the project leader's *skills in leading, coaching, and developing team members*.

The goal of service development is to attract and keep customers who are satisfied, loyal and speak well of the company. Above this they should also be profitable. To really understand customers' needs and wishes, it is appropriate, and often necessary, to involve them in the process of developing and introducing new services. Attractive and customer-friendly services emerge from a dialogue with competent and demanding customers. However, most customers are only familiar with current products and technologies, which means that their ability to guide the development of new services is often limited by their experiences and their ability to imagine and describe new services and innovations. Using a set of techniques sometimes referred to as empathic design (Leonard and Rayport 1997), which is based on observation, might solve this dilemma. By observing customers' behaviour in real situations, the company gets a more objective picture of their real needs and a better understanding of the situations they are 'forced' into and expected to understand and like. The company can learn what circumstances cause people to use the product and how it fits into the customer's system. Furthermore, the company might discover that customers reinvent or redesign the product or service to serve their own purposes and become indifferent to the peripheral or intangible attributes of the product.

In David Silverman's book '*Qualitative Research: Theory, Method and Practice*', Christian Heath (1997) argues that the use of video in the analysis of activities in face to face interactions provides researchers with rare access to social actions and activities. The method permits the researcher to acknowledge both visual and vocal aspects of human interaction along with aspects of the physical environment. The method also provides a complement to the traditional ones and allows the researcher to 'develop insights into the social and interactional organisation of activities and the range of resources on which they rely in their accomplishment' (p. 198).

The relevance of using this type of method can be found in the following quote: 'There are three kinds of companies; those that simply ask customers what they want and end up as perpetual followers; those that succeed – for a time – in pushing customers in directions they do not want to go; and those that lead customers where they want to go before customers know it themselves' (Hamel and Prahalad 1991, p. 85). This chapter provides an example of how SAS, with the assistance of the consultant firm Doblin Group, used observations as the primary source of information in order to find out how passengers behave in the travel process.

Joint efforts by, for example, Wilhelmsson and Edvardsson (1994), Edvardsson and Mattsson (1992) and Norling *et al.* (1992) have resulted

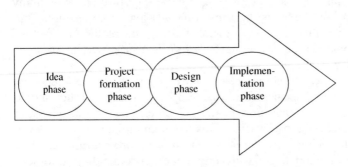

Figure 9.1 The service development process in different phases

Source: Wilhelmsson and Edvardsson 1994. Reprinted with the kind permission of the authors, and of Karlstad University.

in an attempt to set up a model for service development, which consists of three concepts and a process in four phases. The three concepts necessary to describe a service are the *service concept*, the *service system*, and the *service process*. The *service concept* represents a description of what the service provider offers to the customer and is thus closely related to the needs that the service is expected to meet. The *service system*, on the other hand, represents the resources and organisational structure required to produce the actual service. Finally, the *service process* is a description of how the service is, or will be, performed. An important part of the service process is the customer process, in other words, the service process viewed through the eyes of the customer. The three concepts *service concept*, *service system* and *service process* are put into a larger context by connecting them to the phases of a suggested service development process (Wilhelmsson and Edvardsson 1994). Based on empirical evidence Wilhelmsson and Edvardsson (1994) have divided the service development process into four somewhat overlapping phases as seen in Figure 9.1.

The four phases are the *idea phase*, the *project formation phase*, the *design phase* and finally the *implementation phase*. During the idea phase the service idea is identified and evaluated against the current company business objectives. The idea phase ends with a decision, often based on preliminary market research and customer analysis, whether or not it is worthwhile and of interest to the company to further pursue the idea. Once the decision has been made to proceed, the next phase in the process is begun. The objective of the project formation phase is to create a team of people with knowledge and competence suitable for the specific service to be developed. When such a team has been formed the project enters the design phase. It is during this phase that the three concepts discussed above are combined to eventually form the basis for the actual service. The service development process is concluded with the implementation phase during which the service is launched both internally, for example in the form of education and training of employees, and externally with a market entrance.

The SAS case

For a long time the airline industry has found itself struggling not just in terms of profitability, but also in terms of customer satisfaction and loyalty. The industry has been dominated by subsidised national monopolies and suffering from over-capacity. Despite what is often portrayed in commercial advertising, airline travel has become utilitarian, uninspired and outright customer unfriendly. Passengers have been forced to accept this general deterioration and have quietly watched as the airline industry has drifted from its purpose of serving customers. A word better than most others used to describe air travel of today is *fragmentation*. Many customers feel that they are forced into a system characterised by contradictions, redundant or insufficient information, misguided authority and confusion. In this system, they are expected to carry out or participate in a series of activities, each of which individually might seem logical but together may lead to an impression of chaos. The system is also dominated by conditions out of the passengers' reach. Such conditions often seem random and can be overwhelming even to the most experienced travellers.

The background to the case study lies partly in the market trend and effect model presented in Figure 9.2. All arrows point to a time axis. If the different arrows are followed from the origin, they indicate what SAS believes will happen in different areas over time. The model is best interpreted if the customer demand trends are studied first. Changes in these trends will affect the other areas. The effects can be seen following the arches around the figure. The model illustrates how SAS views the development of their customers' demands and how SAS must change in the areas of customer options, supplier relations, organisation and management style in order to adapt to these changes. Consequently, the model can also be seen as SAS's future strategy where the goal is to reach the outermost arch in the figure.

The airline business used to be a mass market with mostly standardised services, and in Sweden a monopoly-like situation for SAS. All customers were offered the same service and treated in the same way and most airlines acted in almost the same manner. Consequently, customers had limited options. Furthermore, different airlines could have a pure business relation with their suppliers and simply pick the lowest bidders as partners. The organisations tended to be very hierarchical with multiple layers of managers. All this forced managers to act as observers. When something crucial happened, it was often too late to act. This way of managing caused an entire industry to merely satisfy its customers, but not to delight or excite them. The airline industry is now long overdue for reinvention. For a long time, it has found itself struggling to keep its head above water in terms of customer satisfaction, loyalty and profitability.

To solve these problems the airline business must aim at fulfilling individual customer needs or even reaching beyond these. One example of how SAS tries to improve its service is the introduction of non-smoking flights. On these, SAS offers smokers Nicorette chewing gum to help them overcome their nicotine needs during the flight. By this simple measure, SAS tries to do more than the customers actually expect. The present challenge for SAS is to come up with a larger number of product variations and to supply the customers with flexible services that have only few limits. In this way, SAS tries to provide services that will allow customers to design and tailor their own travel experience. This in turn will put a lot of pressure on the SAS organisation and its way of working. SAS needs to develop new service processes and a service structure that contains and supports a number of different and changing customer processes. For the supplier relations, SAS will have to think in terms of allied production, meaning not only working closely together, but also relying on suppliers to invent and deliver superior value to SAS customers.

In terms of leadership, the managers must not only be team-leaders (as opposed to 'dictators') making sure that everyone is allowed to do their best, but they must also be active in society. Examples of the latter could be to help in the construction of hospitals or schools, organise education or finance research. Being active in society can greatly impact the image of the company.

Managers also have a responsibility in terms of providing appropriate tools for the personnel in their work to support the customer through his travel process. Using such tools and providing more and better information and training is important and a prerequisite if SAS wants to provide pleasant travel conditions for their customers and give them value for the money. Allowing people to do their best is often a question of empowerment or giving people greater opportunities to make their own decisions. By empowering the staff, SAS stands a better chance of satisfying its customers and even recovering dissatisfied customers. The term empowerment is also appropriate to customers. In other service industries it has become popular to let customers take care of routine tasks themselves. The bank industry is a good example: bank customers can withdraw money from their accounts, move money between accounts or pay their bills at any time without the help of bank staff. This trend is also applicable to the airline industry.

Initiating a new development project

Realising the possible gains for the company that succeeds in reformulating the fundamental concept of air travel, SAS made a strategic decision to become a forerunner in the development of the airline industry.

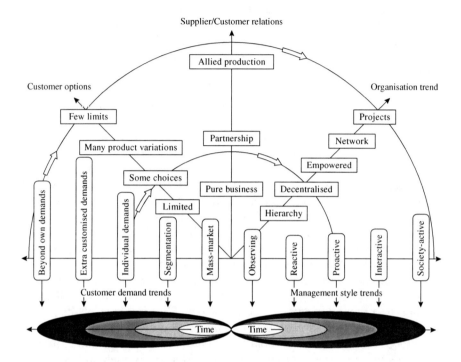

Figure 9.2 SAS strategy and view of market development and trends

Therefore, SAS turned to Doblin Group, a well-known Chicago-based consultant firm, and initiated a project in order to develop a user-centred strategy to prepare SAS for deregulation.

In order to collect relevant data on passenger behaviour, a project group was formed consisting of representatives from top management, SAS product development organisation, and a cross-functional team with both frontline and corporate participants from technical services, marketing, flight and ground personnel. It also contained external design and advertising resources and external marketing resources with social science and anthropological skills. This can be characterised as being a multi-team, i.e. more demanding than a cross-functional, because experts – not just personnel – from different areas were included. In the formulation phase different workshops to help analyse the material were planned. Mostly executives and front-line staff were involved in these workshops. Another round of interviews was also planned.

The aim of the project was to come up with a new product development approach based on understanding passengers' real needs. The project focused on four different areas: *understanding passengers, innovation, personality* and *implementation*. A number of questions were set up around these four areas. *Understanding passengers* involved issues such as: what shapes passengers' expectations? What activities are passengers required to perform? What are the activities they would rather be doing? *Innovations* included questions such as: what industry benchmarks should be surpassed? What innovation systems are required to make travel passenger-centred? *Personality* concerned what role the brand should play. How do passengers conceive the SAS brand name, and how can it be made more enticing? And finally, *implementation* included questions such as: how does a functionally structured organisation realign itself around passenger processes? And how can SAS put together the teams to do this?

The most essential step in turning this strategy into reality was to develop a thorough understanding of the true, spoken and unspoken, needs and expectations of the customers. Further, it was necessary to understand how the customers interact with SAS while travelling.

One of the most dangerous assumptions that can be made by a company is that customers are well aware of their future needs, and that market research in the form of just asking the customers can be used to extract this information from them. Customers can only be expected to know about what is presently on the market. However, they are excellent in helping a company evaluate a new concept. One example of this is the gate cafe introduced by SAS in 1996, in which customers can select their own food and drink before boarding. The customers did not request a gate cafe, but now that it has been introduced it is extremely well received.

A well accepted model that explains what information can be expected to be obtained from the customers is the Kano model of Quality (Kano *et al.*, 1984). The Kano model states that there are three levels of needs: basic, performance and excitement as seen in Figure 9.3. The *basic needs* are the things that the customers expect from the product, for example that it is safe and trouble-free. Because the basic needs are expected, customers are less likely to tell you about them, but their absence is very dissatisfying since they are expected. The basic needs are also low on the satisfaction scale because they are anticipated. The *performance needs* are the customer requirements that are expressed and that companies compete with, for example number of flights to different destinations and ample arm and legroom in the airplane. If such needs are fulfilled, customers are content. The *exciting requirements* are items that the producers develop *themselves*. These items can be satisfiers but not dissatifiers since they are not expected. If they are present they can contribute to customer delight. There is also a time factor in the Kano model: customers get used to exciting requirements, and over time they will take them for granted.

SAS has developed a model of their own as shown in Figure 9.4. This model has great similarities with the Kano model. The fundamental or basic needs that SAS has to deal with are safety, getting the luggage to the right place, and punctuality. These requirements are basic and SAS must meet them first before climbing the pyramid. These activities have,

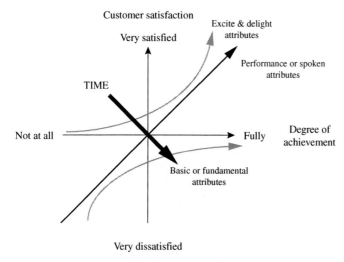

Figure 9.3 The Kano model of quality
Source: Kano *et al.* 1984.

however, been in focus over a long period of time. In order to become more competitive SAS must now concentrate on the activity support, i.e. supporting the customers through the process and help them to do what they want and need to do. SAS must also, in order to delight the customers, tailor the travelling process for each individual.

This cannot be achieved by simply meeting customer requirements. One must probe and seek to understand why something is important, why customers behave in a certain way and why something is perceived as a problem. A company must understand how much of a problem something is and how it may impact on their customers. An environment must be visualised to discover the value that a company can provide. This is essential since it is the delivery of superior customer value that differentiates you from the competition. It is not sufficient to look at the conventional dimensions of competition in your industry. These dimensions are most often well understood by all players. Instead, seek to understand the problems and needs of the customers who use your products and services and create a competitive advantage. Success in customer focused service development requires a deep understanding of:

1 *Customer needs, requirement, expectations and preferences*
2 *Customer service systems,* that is, the technical infrastructure, the customers' knowledge and ability to use services
3 *Customer values and cognitive structure*
4 *Customer's usability processes,* the customers' behaviour when using services. Focus both on what the customer does and what he wants to do!

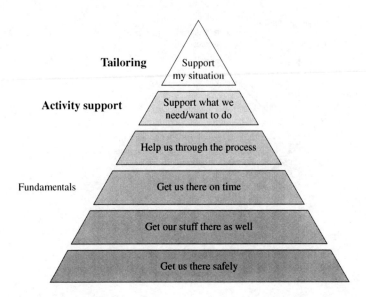

Figure 9.4 SAS has developed a model to understand how to fulfil its customers' basic needs

Source: Powers 1997.

5   *Customers' quality perceptions* such as, is it easy to do business with the company, is the company reliable, and how are customers' dissatisfaction and complaints managed?

Capturing data through observations

In order to answer the above stated questions, SAS and Doblin Group made the decision to study the customers in the environments that comprise the most important elements of the travel process, or, as it will be called here, the travel experience. The travel experience was divided into five phases: check-in, lounge, gate, in-flight and baggage claim, each phase roughly representing a physical location or a function that the passenger has to pass through. Together they include most parts of the interaction between SAS and the customers: see Figure 9.5 below.

Studies of customers' behaviour were carried out partly by using on-site observation, but mostly by using video cameras documenting the customers' travel process. The aim was, instead of being led by rules and regulations as previously, to let customers define their own process. Video cameras were set up in different locations throughout the travel process. In all, SAS amassed close to 1500 hours of video and 3600 items of photographic data of the customers on the ground and in-flight. The video data was studied in sequences as short as five seconds in addition to the still

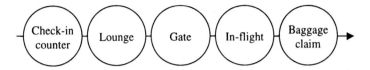

Figure 9.5 The travel experience as a service process

images taken in different situations. This conscientious analysis of data forms the basis for establishing an understanding of the passengers, drawing from actual behaviour and how customers perceive different activities. Furthermore, five weeks of in-depth interviews were carried out with unions, in-flight staff, and ground staff.

Reflection and analysis

The next step in the process was to analyse the different passengers' behaviour shown in the data. This part was done mainly by the Doblin Group. For each of the five phases of the travel experience, both voluntary customer activities and those prescribed by the travel system, were closely examined using the collected video data. By the study, SAS wanted to form an understanding of the true needs and expectations of the customers and use it as a foundation to develop a new service process, allowing customers to design and tailor their own travel experience from a number of options. The analysis was conducted in four steps. First, the video data was divided into short sequences. Second, from the sequences, a number of categories were identified. Third, the categories were further divided into subcategories. And finally, based on the sequences in each subcategory, problems and ongoing activities were identified.

A walk through the halls of an airport quickly reveals that passengers do not engage in any one, focused activity. Whether in line to board, searching for a gate or waiting for a colleague, the sheer number of flight oriented events and the complexity of everyday life suggests an endless list of activities. To study the entire travel experience in detail and then try to develop new products and services based on the individual findings would be impossible due to the multitude of different activities. In addition, a focus on detailed findings would most certainly divert attention from the primary objectives of the passengers. It would be impossible to see the forest for the trees.

However, less obvious to the eye, but apparent through analysis of video data were the larger patterns underlying the myriad of activities. Based on findings from the video data it was possible to find structures that signified and influenced customer activities and behaviour and provided vital information about the passengers. Therefore, the consultants from Doblin Group tried to systematise the different tasks and focus

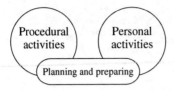

Figure 9.6 Three categories of passenger activities

on clusters of similar customer behaviour. In this way it was possible to identify a number of distinct categories which were divided into three parts: *procedural, personal,* and *planning and preparing activities,* see Figure 9.6.

Procedural activities consist of the mandatory and highly prescribed activities that passengers must perform during the phases of the travel experience. The activities that SAS demands of their customers in order to supply their services are also included here. In spite of the essential nature of these activities, they are often the least understood by the passengers since airlines neglect to provide sufficient information on how to execute them properly. Although numerous options in performing procedural activities are available, customers seldom take advantage of them since they are unaware of their existence.

The procedural activities were further divided into two subcategories, *exchanging value* and *navigating*. Exchanging value represents the system of giving and receiving that takes place as the passengers progress through the travel experience. For example, in exchange for a boarding pass the passenger will be allowed to pass through a guarded doorway to the plane. Often enough, this shuffling of papers is not fully understood by the customers and therefore found unnecessary and bureaucratic. In addition, the customers only have a vague idea of the real value of their tickets and therefore run the risk of losing out on attractive privileges. Navigating represents the activities needed to progress through the many environments and procedures. The video data showed that many customers had great difficulty in figuring out where to go and, perhaps even more important, what to do during different phases of the travel experience.

While procedural activities are induced by system requirements, *personal activities* are instead necessary because everyday concerns must be attended to, even while travelling. Since travelling often takes much time, these activities must find their way in among the dictated ones. Since personal activities are seldom directly travel related, they are often under-supported or even hindered by the airlines. Therefore, personal activities are the most problematic ones for customers to carry out. Doblin Group identified a number of different subcategories of personal activities: *resting, working, entertaining, socialising* and *personal care*.

Resting can range from passengers shutting their eyes for a few seconds to actually sleeping. Long journeys make it necessary to rest, but airport environments are seldom designed to allow passengers to enjoy comfortable and safe relaxation. During long hours of travelling, most passengers try to work in order to make better use of their time. However, the travel environment is seldom conducive to either efficient work or to the necessary privacy. Entertaining represents the activities undertaken by the passengers to pass time while travelling. This involves reading, shopping, playing games, watching TV and so on. Entertaining children, colleagues and business contacts is an important part of these activities. The dominating activity for passing waiting time is socialising with fellow passengers. Entertaining and socialising possibilities together greatly influence the customers' general impression of the travel experience. The airlines, however, often do a poor job of supporting these activities. Finally, personal care represents perhaps the most vital of all personal activities. For instance, to be able to freshen up after a long flight before meeting important business contacts, to enjoy a good meal, or being able to rest are all examples of activities that make the journey more enjoyable. Today, airlines do their best to support customers' personal care, but they seldom succeed in supporting the special needs of the individual passenger.

Since most of the personal activities are not facilitated by the airlines, passengers try to adapt whatever environment and situation they come across to support their needs. Such environments are often designed to support the airlines' opinions about what passengers should do while travelling, which often differs substantially from what customers actually want to do. This reflects a stereotype view of customers rather than genuine understanding of customer needs. For instance, the stereotype picture of business people is that they want to work during the journey. However, if observed, most of them try to relax and enjoy the ride. Another example is the passengers' tendency to 'nest' or 'settle in'. By spreading out their belongings passengers create boundaries against the outside world, shielding off a personal space for themselves in which to rest or work. The question is how does the airline support passengers' needs for enough space for comfortable privacy.

Complying with dictated procedures and at the same time attending to personal issues makes travelling complex and sometimes difficult to manage. Therefore passengers spend considerable time *planning* and *preparing* prior to the journey and how to handle possible unforeseen occurrences throughout the travel experience. Planning and preparing activities were divided into three subcategories: *micro/macro planning, preparing belongings,* and *linking*. Passengers plan their journeys ahead of time as well as while they are underway. When irregularities occur, passengers must often start over and re-plan important parts of their journey, most often without proper resources and support from the airlines. Preparing belongings includes packing and perhaps re-packing whatever it is that passengers bring along when they travel. These activities greatly affect the passengers' abilities to comply with procedures and take care of personal activities.

Table 9.1 Examples of defined guidelines

Personal activities	
Resting	Create environments that allow for different kinds of rest
Working	Provide sufficient workspace Provide information and equipment that facilitate efficient use of travel time Provide resources and support for business travellers
Entertaining	Create different environments and services that help all types of customers to entertain themselves and others
Socialising	Help customers who want to socialise so that they can do this without disturbing others

Passengers try their best to anticipate what will be needed and should be easily accessible in different situations and what can safely be checked in. The most important belongings must be easy to reach, yet hard to steal. Linking includes all the planning and preparing activities aimed at making the travel experience more comfortable.

An abundance of travel information and helpful tips from the airlines support some aspects of planning and preparation. Once the journey is underway, however, the passengers can expect much less support. Delays and changes in flights and connections are almost frighteningly common occurrences which often put passengers in unanticipated situations where they are forced to rely on their own problem-solving abilities.

Defining guidelines and design principles

By identifying the numerous activities that occur throughout the travel experience and mapping the underlying patterns and structures that guide these activities, Doblin Group began to understand the needs and expectations of SAS customers. As a result they were able to define a set of detailed guidelines of new service concepts within each activity cluster. A few examples of these guidelines are given in Table 9.1. The guidelines can serve as checklists when evaluating how new service concepts contribute to customer satisfaction. Finally, they are useful when SAS validates the new service concepts against overall development strategies.

However, based solely on the extensive results from the study, it would be very difficult to choose which development guidelines to prioritise and which new service concepts to pursue. Therefore, SAS needed design principles, which comprised and illustrated on a strategic level what it is that SAS wants to achieve with its development activities. The video data proved to be a priceless source of valuable information, since it allowed identifica-

Figure 9.7 Prioritised design principles

tion of which areas were suffering the most from lack of customer focus. SAS identified three such design principles: give passengers control, make the process transparent, and empower the staff as illustrated in Figure 9.7.

Although all of the activity categories identified in the study are considered equally important to the customers, the study confirmed that services provided today are highly influenced by the company's needs and the rules and regulations dictated by the system. Under these conditions, the customers try to carry out their personal activities but are more often than not hampered in their attempts. This forces passengers to resort to unconventional solutions in order to achieve their own personal objectives. Since the system is focused mainly on procedures and does not allow passengers to interact according to their own personal abilities, people are treated as luggage that is inspected, tagged and sent on its way through the travel experience. It is vital that SAS succeeds in engaging their customers in the travel experience by letting them decide for themselves what level of support they need. SAS must provide the passengers with information that will empower them and allow them to serve themselves through their journey and give them freedom to take active part in creating the travel experience. In this way SAS can give them control over their own experience, feed back and feed forward.

The rules and regulations that dictate the travel experience often give the customers an unnecessarily complex impression. In addition, there is a lack of clear and consistent help to understand how the process works. SAS cannot by itself eliminate all the complexity of a dictated system but is certainly able to help remove the mysticism that traditionally surrounds it. SAS can develop its own processes so that they become as simple and understandable as possible. If the procedural activities are made simpler to understand, learn, and complete at the same time as more support for personal activities is provided, a balance between procedural and personal activities can be created. An important issue will be for SAS to help passengers understand the process. SAS personnel is today often stopped from doing their best in giving the customers as pleasant a travel experience as possible. Information systems and other tools are not designed to give the staff the opportunity to interact with the passengers in an effective manner. The staff needs better tools, more information and better training so that they can assist the customers instead of having to defend a bureaucratic

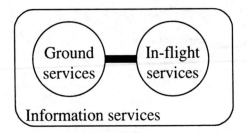

Figure 9.8 Prioritised areas of innovation

system. This would also allow the staff to personalise the services to the individual customer needs without compromising efficiency. SAS must supply its staff with the right tools to create the best opportunities for serving customers. The aim must be to create a service system that stimulates and simplifies both procedural and personal activities.

Based on the findings in the study, the development guidelines, and the design principles, SAS realised they had to adopt a comprehensive view of the travel process instead of focusing on separate parts. As a result, three main areas of innovation were identified: in-flight services, ground services and information services, see Figure 9.8.

Ground services and in-flight services more or less correspond to the different phases of the travel experience, whereas information services serve as the fabric that connects the different phases and binds them together. The aim of focusing on the entire process is to achieve as positive an impact as possible on the customers' overall impressions of the travel experience. In addition, keeping these three areas in mind will allow the creation of innovation portfolios, i.e. series of interlinked innovations, that will not only be very powerful, easy to understand and implement, but also very difficult for competitors to imitate.

Brainstorming and concept development

Doblin Group now turned to the stage of coming up with new service ideas by means of brainstorming. Important as a point of departure were previous problems and ongoing activities identified in the data, the service concept guidelines, and the design principles. Based on this the consultants asked themselves two questions, how can SAS facilitate the customer's situation, and how can SAS solve the problems customers are experiencing.

Brainstorming is an essential part in every innovation process. The purpose is to transform observations into graphic, visual representations of possible solutions. Brainstorming is often considered an undisciplined session in a creative process. This, however, is not true. The sessions are guided by rules, and each participant is asked to defer judgement, use others' ideas to build on, stay focused on the topic, and encouraged to come up with wild ideas. Only one conversation at the time is allowed (Leonard and Rayport 1997).

Doblin Group's brainstorming sessions resulted in 150 innovative ideas. The new ideas ranged from small modifications in the service system to offering new options for the customers in the service process.

Discussion

One of the primary outcomes of this study is a deeper understanding of passengers' needs and concerns throughout the service process, which would lead to re-design of existing services and to the development and implementation of new and innovative services. Most important of all was to find a new approach in developing services that clearly focused on customers' perception of the service process. This new approach will aid SAS in developing not only separate innovative services, but a series of innovations complementing each other to form a holistic travel experience based on the individual needs of the customers in the different episodes during the travel process. This is in line with what Edvardsson (1997) found to be crucial for success when developing new services or re-designing existing service offerings.

Overall, SAS has identified 40 problems that customers encounter which justify drastic product development. The company has also acquired unique information leading to some 50 minor improvements. These innovations need to be properly staged and implemented. They also require develop-ment of communication pieces such as documents and videos.

In order to develop new services it is important to select a method, or preferably a number of methods, which support the service process and system. The case provides an example of how companies can use obser-vations as a means of mapping true customer needs. This provides a new approach for SAS, a company that previously focused mainly on company regulations. By observing customers' behaviour and overall travelling processes, and trying to figure out what people wanted to do during their travelling, SAS let the customers design the process instead of squeezing them into to some company-designed process. This was done as a first step to come up with a larger number of product variations and to supply customers with services that have only few limits. In the future SAS hopes to be able to give customers the control to design and tailor their own travel experience. This however, will put a lot of pressure on the organ-isation and its way of working. SAS must develop a new service process and service structure that contains and supports a number of different customer processes.

SAS identified a number of other benefits from observational research using video cameras. First of all, video recording represented a new tech-nique not used very much earlier for service development. Therefore, aspects not covered by more traditional customer survey techniques were now included for the first time. In addition, the video recordings allowed study of customer behaviour at an extraordinary level of detail, over and over again, without running the risk of losing any vital information along the way. The video recordings included sequences where the passengers solved their own problems, and this information could prove vital when designing new services, especially if the person being recorded could be involved in the interpretation of the sequence. Further, the video data-base in itself constitutes an enormous source of ideas for new and improved products and services. It will be possible to return to it time and time again for new inspiration. Finally, SAS has recognised the video database to be an important part of an ongoing learning process within the company. As such, it must be continuously updated through active contacts with customers and through renewed use of video recordings. In this way, SAS can make sure that they keep the understanding of the passengers' needs and expectations up-to-date.

SAS could have strengthened the findings in this case if they would have let customers take part in the interpretation and analysis phase. We argue for an approach where customers are also involved in one way or the other when data is being analysed. This is based on the idea that the customers' frame-of-reference should also be reflected in this phase of the study. This does not mean that consultants or internal experts in a service company cannot and should not make their own interpretations.

It is a mystery to us why SAS did not use the multi-team formed earlier in the process. The team consisted of representatives from top management, SAS product development organisation, and a cross-functional team with both frontline and corporate participants from technical services, market-ing, flight and ground personnel. It also contained external design and advertising resources and external marketing resources with social science and anthropological skills. This is a textbook example of how a team should be built. SAS should have let this team be more responsible in the analy-ses. In this way the company could have developed its own skills and abil-ities to analyse this type of data and ended up with an even bigger knowledge about the customers. The customers themselves should also have been involved in the data analysis. It might not be that easy for someone else to figure out what is actually happening in a particular sequence. Who better than the customer himself can explain what happened?

The method in this case has been used in other service contexts where it is possible to observe customers' actual behaviour during the service process, but the empirical results are not directly applicable in other service com-panies since they are rather context and company specific. However, the findings do show that new service development will benefit from observing customers' actual behaviour and that data on customers' processes is useful in the development of new and innovative services and in the improvement of existing ones.

Conclusions

This chapter is about the development of new services and service innovation based on the service logic. In our introductory theoretical frame of reference we give an overview of research on service management and the new service development and innovation process. We make no clear distinction between service development and innovation and argue that the general innovation theory is applicable also in service organisations but emphasise that we have to understand the service characteristics and the service logic. The case study shows that by studying or mapping the customers' behaviour, a basis for service development and innovation is created. This is one way of collecting data about the customers' 'real needs' that helps us better understand how to design customer-friendly service offerings and service processes. This is one way of strategically responding to change-processes in society. This is part of what in Chapter 1 is called 'innovation under conditions of reflexivity' and how customer-driven innovations may be managed. Strategic reflexivity is a process that makes it possible for customers and the service provider to adjust and change their roles. In this chapter we have emphasised how data on customers' behaviour may be used to re-design and adjust a service process to customers' real needs and support their value creation processes.

10 Systems of innovation in social services

Lars Fuglsang

Abstract

This chapter seeks to conceptualise the different production systems that underlie innovation and development processes in social services, more specifically home help services for the elderly in Denmark – drawing on an already reported case study.* I distinguish among what I conceive to be three main production systems: a societal system, an interactive system and an organisational system. Each of these systems has developed over a period of ten to fifteen years since the beginning of the 1950s. However, they can also be seen as simultaneous occurrences that exist side by side. A further argument is that each of these production systems develops a specific management role. We will particularly look into the management problems of home help as an organisational system in the latter stage.

Home help in Denmark

The evolution of home help in Denmark can be divided into three stages that each can be characterised as a specific service production system (for a more detailed description see Fuglsang, 2000):

The first stage is the period from the mid-1950s to the mid-1960s where a 'societal' system of home help was developed (cf. below). During the early 1950s, several privately organised home help services were set up (organised, for example, by the Salvation Army or the Church). In addition, a small number of municipalities (Copenhagen, Frederiksberg, Hvidovre, Odense, Esbjerg, Århus and Randers) had, during the period from 1952 to 1956, developed a public home help service. In 1958, an Act was passed in the Danish Parliament (no. 100, 1958) that made it possible for all municipalities to develop a home help service, financed by

*the case study in this chapter is reprinted from the *Scandinavian Journal of Management*, vol. 17, no. 4, L. Fuglsang 'Management problems in welfare services: the role of the "social entrepreneur" in home-help for the elderly, the Valby case', pp. 437–55 © 2001, with permission from Elsevier Science. Cf. Lars Fuglsang (2001).

the state. In 1968, another Act (no. 230, 1968), made home help services obligatory for the municipalities.

The discourse about home help at this stage was a broad socio-economic one. Home help was seen as an economically more efficient solution to the care for the elderly than residential homes. Another intention behind home help was to relieve women of their domestic obligations (taking care of the elderly among other things), in order to attract more women to the labour market. Home help could be a more acceptable and less dramatic solution for the family than the residential homes.

There was no educational requirement for home helpers in the 1958 amendment. Home help was considered to be a simple task, even simpler than housewife relief, a similar arrangement (created in 1949) aiming at replacing housewives in cases of illness or childbirth.

Home help was carried out by dedicated and experienced women who knew what to do from their own experience as 'housewives'. Furthermore, they worked on an individual basis without seeing much of one another. They were formally managed through a central office which assigned them a weekly timetable. Practical problems during the day and changes in the weekly timetable were co-ordinated over the telephone. The home helpers very often developed a strong personal relationship with their clients. A home helper might refer to a client as 'my' client. The personal relationship between a home helper and her client regulated the content of the service to a large extent. There was time for coffee and small-talk, which was sometimes seen as more important than ordinary tasks such as cleaning. Class differences between home helpers and their clients were sometimes important in the relationship. In upper-class homes, home helpers sometimes felt they were treated like servants.

The production system at this stage may thus be termed a societal system since it to a large extent relied on broad expectations to the social role of the housewife and the very broad socio-economic discourse about home help. No intermediary systems and interaction among home helpers was present in this stage not to speak of a well-developed organisational system within the municipality.

The second stage covers the period from the mid-1960s to the end of the 1980s. The 1974 Act on Social Security (no. 333, 1974) was particularly important, in addition to the above mentioned 1968 Act.

Following the 1968 Act, the municipalities were *instructed*, for the first time, to establish housewife relief and home help, which thus became obligatory for the municipalities. The 1974 Act removed the concept of 'housewife relief', which was replaced by the term 'temporary home help', to be distinguished from 'permanent home help'. Furthermore, for the first time, home helpers had to be qualified. The requirement was later specified in terms of having undertaken a seven-week course. As a result of these changes the voluntary organisations disappeared from the scene, and home helpers started to see themselves as semi-professionals rather than as 'housewives'.

During this stage the policy known as 'as long as possible in your own home' started to have its effects. State reimbursement to old people's homes was abolished in 1973, but reimbursement for residential homes for the very weak elderly was maintained. Many more elderly people, including the weaker ones, now stayed in their own homes. Home help therefore became more challenging and difficult, requiring a minimum of education and experience.

In a written report from a Commission on the Elderly, set up by the government, a new principle of care for the elderly was established (Ældrekommisionen, 1980). Care for the elderly was to be seen as a professional activity, a kind of 'treatment' for the elderly, making it possible for people to maintain their 'social roles' as long as possible. Services for the elderly were to be regarded as a kind of socio-psychological treatment. 'Activation' was another key word in this period. For example, home helpers should seek to activate the elderly, according to a circular from the Ministry of Social Affairs, 'Instruction for the home helper' (no. 208, 1975). Thus, home help began to be viewed with greater respect and as something that could become a true profession rather than a simple job.

During this period, some experiments with the collective management of home help were carried out. Home helpers were organised into districts; transportation costs were thereby reduced and the replacement of home helpers in cases of illness could be organised more smoothly. In some municipalities, home helpers were collected into self-managing groups, each with responsibility for a district. Collective management generally promoted a feeling of growing professionalism among the home helpers.

The production system developed during this stage is an interactive one because various rooms for interaction 'behind the scenes' were created. More precisely, a borderline was drawn between the front stage of the service, in the home with home help during the actual work of the individual home helper, and back stage, training and planning activities among colleagues in office buildings or schools. However, stronger organisational routines did not develop at this stage. It was rather an interaction among colleagues at the personal level that developed. The home helpers themselves would, through their mutual interaction, create the home help service and expectations to it.

The third stage began in the early 1990s. Some elements belonging to this period date back even earlier. The financing of home help was changed in 1985 (Act no. 262, 1985), when the government's 50 per cent reimbursement of home helpers' salaries to the municipality was removed. Municipalities were now to finance home help through block grants from the government. This was supposed to strengthen economic responsibility in the municipalities. Further, the 1987 Act on Accommodation for Elderly and Handicapped Persons (no. 378, 1987) stipulated that residential homes for the weak elderly could no longer be built. Instead, the municipality could build individual homes for the elderly, for which they could provide home help. Services and housing were thus separated.

These changes meant that it became easier for the municipality to think of home help in terms of a service organisation that provided home help within a district in an efficient way. Integrated collective home-care, combining home-nursing and home help, started to evolve, especially after two new collective bargaining agreements were reached between the municipalities and the nurses' and the home helpers' organisations in 1989 and 1991. These promoted employment at the district level.

During this period, culminating in the 1997 Act on Social Services (no. 454, 1997), the procedure for the allocation of home help started to become more systematic, with a legal requirement for written consideration of cases (Act no. 1114, 1995). and attempts by the municipalities to create a more standardised 'language' for dealing with home help services. In 1995 (same Act as above) elected committees of the elderly (ældreråd) and committees taking care of complaints (klageråd) became obligatory for the municipalities. Several municipalities started to distinguish explicitly between the 'ordering' of services (allocation procedures) and the 'carrying out' of services. Furthermore, the ministry of social affairs requires of the municipality that it once a year must create service standards for personal and practical help and also develop service declarations.

A further characteristic of this period was a change in the training of home helpers. According to the above mentioned law of 1995, a qualification was no longer obligatory. This was partly due to persistent recruitment problems in the home help services. However, a broader and more visible training scheme for home helpers was created in 1990 (Act no. 423, 1990), and was widely used by the municipalities. While giving home helpers a broader, more theoretical background, this training scheme also recruited trainees from a social group consisting of young women with low self-esteem, women who had often failed at school or in social life. They considered themselves as 'wage earners' rather than housewives or semi-professionals.

During the 1990s, a more professional and technically managed organisation was thus emerging, which, among other things, could cope with the accountability problems of the welfare state and restore the 'social contract' with society. There was a shift of attention away from semi-professional management and towards technical and organisational management. The elderly client was increasingly being regarded as co-responsible for his or her situation, rather than as someone undergoing treatment. The client started to have a relationship with a service organisation rather than with an individual home helper.

Thus, in this stage, the previously mentioned differentiation of front stage and back stage came into focus. Back stage activities were put into question and could no longer play such a dominant role. Home helpers, in fulfilling their roles, increasingly had to take into consideration a number of external demands (economic, legal, political). This was possible only if home help considered itself, and developed into, an organisational system that was more strongly programmed in relation to the external demands. This in turn required a new form of 'strategic reflexivity' (cf. the introductory chapter) to bridge the gap between external demands and internal resources and job-motivation. Each local home help organisation had to interpret its potential role more actively in the division of labour and in service development more generally in order to survive.

The story of home help in Denmark can thus be described in terms of three production systems that succeeded one another: a societal system; home help was regulated through the social role of the housewife and a socio-economic discourse concerning the value of home help. An interactive system; a borderline between front stage and back stage activities was drawn; home help became more regulated by back stage activities, i.e. interactions among employees. In addition, an organisational system; home help developed into an organisational system that was more strongly programmed in relation to external demands and obligations.

Each system involved a new actor role in home help with a different attitude to their work and different strategic behaviour. In the first system, home help services were carried out by self-motivated and experienced housewives without training and managed by an administrative case officer. In the second system, home help was carried out by 'semi-professionals' with a seven-week course behind them. They worked in groups, mostly on a full time basis, and were supposed to 'activate' the elderly. Management was carried out by the group as a whole. Moreover, the experienced home helper was assigned the role of inspector (*tilsynsførende*). Often she acted as a 'mother' to the employees. In the third system, home help started to be carried out by a group of 'wage earners', young women who could earn their incomes in this way. They often had no special personal or professional preference for this job, but they were usually better trained than before. In this stage, new management roles were created with a stronger emphasis on strategic reflexivity, as we shall discuss below.

Home help in Valby

Valby is a suburb of Copenhagen with about 44,000 inhabitants. Recent figures from the Valby website show that 2,350 persons receive home help in Valby, including 62 per cent of all women over the age of 80, 22 per cent of all women between 67 and 79, 33 per cent of all men over 80, and 15 per cent of all men between 67 and 79.

The story of home help in Valby must be considered in the light of the following background. From January 1997 to December 2001, Valby has participated in a pilot project on local government in Copenhagen together with three other local authorities. The purpose of this experiment was to strengthen democracy, improve municipal services and streamline municipal administration.

During this period Valby had its own local council with 21 members. Direct elections were held on 21 May 1996, and as of January 1997 Copenhagen City Council transferred powers to the local council in several areas according to regulations adopted by the Copenhagen City Council on 28 March 1996. Nine committees were set up, including a committee on the elderly, with responsibility, among other things, for home help services in Valby.

The experiment with the local council changed the context of home help in Valby in several related ways. For one thing, home help became linked to a local democracy project. Politicians and administrators (whose jobs could depend on the outcome of the experiment) had to take initiatives that could strengthen the relations between the municipality and the local population.

Several initiatives were taken to improve communications between the municipality and the members of the public. In the local newspaper, people were informed of the initiatives and meetings of the local council. A 'citizens' shop' was opened near the new town hall where citizens, among other things, could obtain information about public services in the community. Furthermore, an interactive homepage was also established, but was not very successful. By August 1999, only fifteen contributions from members of the public and politicians had been published on the website, which was started up in November 1997.

The municipality also had to be able to demonstrate to the Copenhagen City Council that services were improving and that administration had become more streamlined. Initiatives in these areas would have to be different from initiatives in the city of Copenhagen, if it was to be possible to distinguish between the two. The local council was dependent on evaluations of the projects, and on visible and measurable results within a relatively short period of five years.

The core of the local council project could be regarded as the ability of the municipality to come closer to the people, which calls for greater interaction between the two. This applies to services as well as political decision-making. In the so-called 'value-base' for home help in Valby it was stated that 'good work' should take its starting point in the individual human being. Work should be characterised by interaction between employees of the service and their clients, as well as by respect, tolerance, openness, confidence and solidarity. Employees should generally look at people as individuals, with all their differences, communicating openly and with respect for the dignity of the individual and ensure that the integrity is maintained.

In the broader societal context, the Valby approach to the problems of the accountability of the welfare state, as discussed in the previous section, consisted of closeness and interaction. This strategy was somewhat different from the usual one, with its greater focus on the standardisation of services. The closeness strategy was reflected, among other things, in the way management problems in the home help services were perceived. I shall be discussing this further below. Interviews conducted in Valby, together with various documents handed to me to complement the interviews, tell the following story about home help in the community:

Home help was originally managed centrally by the municipality of Copenhagen. During the 1970s, it was decentralised to local health insurance offices, while at the same time a new semi-professional organisational structure started to evolve. Until the passing of the Act on Social Security in 1974, home help could not be granted without a requisition from a doctor or a midwife, and nurses then assessed citizens' need for help.

With the Social Security Act, a requisition from a doctor or a midwife was no longer needed. A new semi-professional role was established: the inspector (tilsynsførende). The inspector's job was to assess the clients' need for help, to inspect work in clients' homes and to organise the daily work of the home helpers. Special administrative case officers managed the granting of home help together with personnel management and resource management. The inspector was typically an experienced home helper who had taken the regular seven-week home helper course, as well as various supplementary courses and a special seven-week inspector's course.

In 1988, however, the Valby home help scheme was integrated with home nursing (regulated under a separate act). Professional demarcation problems followed. 'Section leaders' were now employed to supervise the work of home nurses, inspectors and home helpers, who were all grouped together in 'sections'. The administrative case officers thus became less important. Many case officers left or were transferred to other jobs. The section leaders and inspectors took over administrative tasks, leaving less time for the inspectors to visit clients and inspect the work in the home. At the beginning of the 1990s, when collective home help was introduced, the inspectors were also supposed to supervise the groups, although they had no formal management powers. The inspector was in practice a kind of 'foreperson'.

From 1994 onwards, home help in Valby was organised according to a 'framework plan' applying to the whole of the Copenhagen City Council area. It was placed within a Department of Care. At the top of the department was a chief social worker (socialfaglig leder), and under him were the section leaders as before. This framework plan set the stage for collective, semi-professional management of home help in the city of Copenhagen (at a rather late stage, that is, than in the rest of Denmark). According to the plan, the 'home helpers were co-responsible for dealing on an everyday basis with the work assigned to the home-care group. The inspector was to participate as much as possible in the meetings of the home help group'. The home helpers were allowed to meet for fifteen minutes every morning and for half an hour during the day after lunch at a daily 'gathering' – on condition that this gathering provided professional input and did not reduce the total output of the group.

The Valby Department of Care was divided into three sub-departments (plus a training department and a department for aids and appliances), each with 2 sections and section leaders, 10–13 home nurses, 100–120 home helpers and 2–4 administrative workers. The home helpers were organised in groups of about 14 with 1 inspector covering 2 groups.

Changes in the home help system took place in 1998–9, and were prepared in 1997 during the local council's first year of existence. In 1997, the Copenhagen City Council had produced a report on sickness absence, showing that sickness absence in the council area as a whole was 18.9 days per year on average. In the home help services, however, it was as high as 34.4 days a year. These figures were discussed intensively by the Valby administration officers and councillors. According to an internal note from the Department of Care to the Valby local council, sickness absence in home help services in Valby in 1997 was 38 days per year on average. This was of course extremely high, and unacceptable. It posed a strategic challenge to the new local council, both as an accountability problem and as an indicator of low job satisfaction.

In the Department of Care, interdisciplinary groups were set up in April 1997 in all three sub-departments, with the purpose of discussing thematic issues related to the home help system. Employees and managers both took part in the discussions in these groups. From May 1997 on they started to work on the problem of sickness absence.

The work of these groups was co-ordinated by a steering committee. This included the chief administrator of Valby, the chief social worker, representatives of employee organisations, representatives of security committees and representatives from each interdisciplinary group.

The steering committee came up with the idea of *group management (grup-peledelse)* as a possible solution to the problems. The chief social worker was a driving force behind this proposal. A report on group management was prepared and was submitted to the sub-department groups in November/December 1997, where it was well received. An application was then drawn up and submitted to the Ministry of Social Affairs. The Ministry approved the application in May 1999. The project had already been adopted by the local council in 1998 however, and was initiated in December 1998.

The basic idea was to *replace the existing 14 inspectors with 21 group managers (later reduced to 18)* and to give them a higher salary and formal powers to lead and distribute work among the home helpers, *thus removing the semi-professional, collectively managed home help scheme*. Thus, it was considered that collective home help management was not working properly, and was, at least indirectly, identified as a major reason for the unacceptably high levels of sickness absence.

To become a group manager, two and a half years' training as a social and health assistant (*social- og sundhedsassistent*) at the new training college for social workers was required. The inspectors already employed in Valby, however, were offered the chance to upgrade themselves, while still enjoying full salary. By the summer of 1999, two inspectors were undergoing the additional training required. Five new social and health assistants had been recruited as group managers; three inspectors had been transferred to the new positions because they had already completed the relevant training, while five had been permitted to continue on grounds of other personal qualifications.

The job of group manager was described in the application to the Ministry of Social Affairs and other places as follows. (1) The group manager is a 'cultural' leader paying attention to values and attitudes. Maintaining a dialogue with employees on these issues is seen as important. (2) The group manager is a personnel manager who closely follows the activities in the group. He or she is also to conduct individual conversations with all employees (*medarbejderudviklingssamtaler*) and talk with employees individually in cases of sickness absence (*sygefraværssamtaler*). (3) The group manager is an administrator. He or she should focus on long-term planning. The group manager formally directs the group and has the power to decide which home helpers serve which homes. (4) The group manager must spend time in the field. He or she must be visible to the client and the home helper in the clients' homes. He or she must supervise and advise the home helpers. The group manager must spend at least half the time in the field. (5) The group manager must communicate with the clients in order to strengthen their trust in the organisation. He or she must give the impression that the home help scheme is organised for the sake of the client, but must also be able to set priorities.

From tape-recorded interviews with section leaders, group managers and home helpers in Valby it was possible to identify at least two people who were regarded as successful group managers. They had been able to reduce sickness absence in their groups and had spent time in the field. These group managers more or less shared the following biography (although they had quite different personalities).

They were dynamic women in their mid-thirties, who had recently been recruited to Valby. Before coming to Valby they had worked and gained experience in several other home-care organisations. One of them had worked in other related areas (and still did so in her spare time). This career pattern differed in a general way from that of the upgraded inspectors most of whom had worked in the same organisation for many years. These two group managers had both made a conscious choice of home care as their preferred working area. They had both been trained as social and health assistants. Both emphasised the virtues of this education in terms of the professional competence it ensured (almost comparable to that of a nurse).

Both were popular in their group and ready to acknowledge this. They also willingly accepted the role of manager with its formal powers, something that was more difficult for the upgraded inspectors. They were both committed to group management as a concept and loyal to their superiors. Caring about employees was a crucial element of their job. They were able to promote home help as an interesting and beneficial job in their contact with the home helpers (according to the home helpers themselves), focusing, among other things, on the social value of the job and its variety within the day. By contrast, co-operation with other group managers was regarded as difficult. Further, they both thought there was too much bureaucracy. Both gave the impression of being overworked and a bit frustrated by the many problems facing group management, such as a computer system for Copenhagen ('TOP') that did not suit Valby.

The group manager, according to other respondents, is a person who can act as a 'mother' or a 'parent' to the home helpers. This concept of group management was rejected by top management, however. The group manager is a co-ordinator, rather than a mother. Professional qualifications are important. However, personal qualifications are important too. The group manager must be a practical, competent person able to communicate at various levels – professional, personal and technical. He or she should be able to 'walk around' and talk to colleagues, and act immediately if something goes wrong.

Explaining the new management role

The management problem in the Valby home help scheme can be conceptualised as tensions between three potentially conflicting 'service relations', namely the personal, professional and technical relations. They are each linked to one of the three different previously discussed production systems.

Personal relations are the bonds that exist between a specific employee and a specific client. The relation is normally very important to in-person services in general, and to services for the elderly in particular, because of the repeated personal contact between the employee and the client. The personal relation exists if the service worker can say that she has a client, an elderly person, whom she would call 'her own'. This relationship is dominant in the societal production system regulated through the social role of the housewife. The personal relation implies that users and producers, in fulfilling a role, come to share a common history. This can be an incentive to co-operation. Since they have to see each other 'next time', it will be an advantage for each of them to co-operate (cf. Robert Axelrod's co-operative game theory). For these reasons, personal relations are often reported as more satisfactory than anonymous relations. There is a feeling that the service can more easily be anticipated and influenced by both the worker and the client. The negative side is that personal relations may be difficult and costly to establish. Furthermore, once they are created, they are not so easy to discard, should they develop in an unsatisfactory direction.

Professional relations are also very important to in-person services, including services for the elderly. The professional community is of course important in terms of pay and general working conditions. Nevertheless, professional communities also often maintain a discourse about good practice and theory, as we have seen. They are interactive systems that provide a professional identity for the service worker. An example is the professional role concerned with the 'activation' of the elderly. Professional identity can be a protection against pressure from both the management and the client. The professionals often share a history and many experiences with colleagues, and these constitute bonds of loyalty and establish structures of motivation. However, professional communities and identities may not be sensitive to external demands as previously discussed. They are flexible and adaptable to overall theoretical and professional developments, general discussions of professional principles etc, but they are often more insensitive to the needs of the individual client or consumer groups.

Technical relations are those that the worker has with other workers or clients in the practical division of labour. Relations among employees are not only regulated by professional rules and routines; they also have to be organised in a more practical sense on a daily basis according to specific circumstances and demands. In the case of in-person services for the elderly, technical relations often have a separate space and time set aside during the day. They cannot be dealt with during the job itself, since this is individualised (individual home helpers working in the private homes of the elderly). The distribution of tasks and the division of labour, for example, are organised at separate meetings during the day. This can be arranged in different ways, depending on the organisational style or the task in question. For example, the work may be organised by a manager, by a core of skilled workers (nurses) or by the whole group of service workers. The practical distribution of tasks does not always fit the professional ideals, for example in situations with high sickness absence. The technical aspects of work become more important when home help develops into an organisational system that is more strongly programmed in relation to external demands.

In the story of home help in Denmark we have seen how these different aspects of the work have been emphasised at different stages in history, although they may have been present simultaneously all the time. However, different groups of home helpers tend to be inspired by different aspects of the work, depending on the stage at which they first came to be employed in the service. For example, the young women (the 'wage earners') today recruited direct from school, appear to be sensitive primarily to the technical aspects of the work. They want to know what to do, and how. Home helpers employed during the late 1980s (the 'semi-professionals') would form a group mainly concerned with semi-professional aspects. Moreover, those recruited 20 to 30 years ago (the housewives) are inspired mainly by the personal relationships with 'their' clients. A fourth group, nurses and section leaders, has a strong sense of professional identity and 'public service', while also having the power and authority to decide upon many aspects of the home help activity.

Thus in the Valby scheme *the group manager is a 'social entrepreneur'*, by which I mean an entrepreneurial person who breaks down these fossilised social/professional relations, and facilitates the creation of new ones. This person is not a conventional 'team manager', whose job it is to motivate employees. Rather, her job is to confront people with various 'unpleasant' aspects of the work and persuade them to assume responsibility for such things, perhaps embarking on some kind of *bartering*. She introduces 'strategic reflexivity' in work by helping the employees to consider their role in relation to external demands and the organisational and strategic programming of them.

Further training, courses, new clients, new job functions, variety in the course of the day, small social events etc. can be offered in such a strategic reflexive process. In order to enhance on strategic reflexivity, the social entrepreneur must be focused on concrete organisational problems and should not be tied up with either specific social roles or professional inter-action systems. She must be able to break the momentum of professional or personal relationships, if or when these become problematic. If something goes wrong, she should not be afraid to interfere. Simultaneously, she looks after her colleagues, knows their various personal data by heart (birthdays etc.), and can talk to colleagues and clients during difficult times. The social entrepreneur is not greatly interested in general professional and political issues. The day-to-day organisational problems and improvement of the spe-cific working conditions are what should interest this person. She should try to handle the organisational dilemma between the need for organisational programming on the one hand and individual motivation on the other. She should try to find the right people to the right job and as such, she is a co-ordinator. In this, she fulfils a *function*; her position is not a *privileged* role.

There are several difficulties attaching to this role. A person in such a position has to change the dominant image of management as being either bureaucracy or 'foremanship'. She must transform it into a more sophis-ticated role that possesses both 'pros' and 'cons'. She must also be able to conduct conversations with individual employees and differentiate between them, rather than addressing herself to a collective group. Furthermore, she must explicitly be prepared to use her formal powers, which may be difficult if she is regarded as a colleague, 'one of us'. The typical social entrepreneur is a woman, but there may be some conven-tions and codes that make it difficult for women to undertake the role. One last problem for the role of the social entrepreneur may be that it differs in some ways from that of the classic (highly individualistic) entre-preneur, since it is subject to certain restrictions, such as the required loyalty to top management. Thus, if a 'classic' type of entrepreneur is engaged for the social entrepreneur job, conflict may easily arise between top management and this person.

How can support for this role be institutionalised? In Valby, the following options have been suggested. For one thing, training has been seen as an important factor. The social entrepreneur must have a broad professional background in order to be sensitive to the employees' needs and problems, and must have real authority. A relatively high salary is also required to attract and keep the social entrepreneur. The (limited) data from Valby suggest that the social entrepreneur will tend to be mobile on the labour market, and therefore difficult to keep. Furthermore, she must be able to offer something to the employees in the bartering process discussed above, and must therefore have some kind of budgetary control (even if it is limited), or access to other types of benefit. Continuous encouragement and support from top management is important. In particular, women who have been employed in the position of inspector or something similar may sometimes feel uncomfortable in the new role for reasons of convention, and may need continuous supervision. Brainstorming meetings with group managers is one solution that has been tried in Valby. Such meetings should probably not be made obligatory, since they might appear rather irritating to efficiently functioning social entrepreneurs. Group managers should probably also be allowed to build up their own networks, internally and externally. For example, the social entrepreneur should be able to attend seminars and courses of her own choice, both as a personal benefit and as a source of inspiration. Finally, the recruitment of social entrepreneurs should be better organised. It is crucial for the organisation to have a recruitment strategy, and to know what type of person it is looking for. Giving 'social entrepreneurs' a higher salary would also make it possible to employ them for limited periods.

Is flexibility a feasible strategy in in-person services in general?

Is it possible to combine an organisational production system of the above kind with flexible service production? Often this kind of development is described as a development towards standardisation of services. But in Valby, it was claimed that the new production system will imply more flexibility *vis-à-vis* the client. Is this a feasible strategy?

The answer to the question is a cautious 'yes'. Five reasons are listed below, implying that flexibility could theoretically be regarded as a feasible choice for service organisations in general. They are: (1) the support to be found in the literature on services for a 'second divide'; (2) the productivity problem characteristic of in-person services; (3) the substitution effect; (4) the fragmentation of services; and (5) the apparent job and client satisfaction associated with flexible specialisation.

The support we find for a 'second divide' in the literature on services

From this literature as a whole, with its various perspectives on service development, we get evidence of a 'divide' in services between standardised and flexible production (cf. also the literature on flexible specialisation, for example Hirst and Zeitlin, 1991). To take a few examples, on the one hand there is a body of literature in which it is argued that services are produced in much the same way as manufactured goods, and that the assembly-line principle and standardised production can be applied to services, something which neatly reflects the growing 'industrialisation' of services (e.g. Levitt, 1972). Attempts have also been made to apply the idea of 'lean production' to services (e.g. Bowen and Youngdahl, 1998). Lean production is a more flexible form of the assembly line system. On the other hand, there is other literature that stresses the particularity of services. It argues that services are usually produced as they are consumed, and that they are difficult to store. You never really know what you are getting. Thus, service products are usually characterised by a high degree of simultaneity, perishability and/or intangibility. Services are therefore necessarily flexible (for a review of these arguments see Sundbo, 1998a). Some 'flexible standardisation' may take place nonetheless. A service can be 'modularised', meaning that the customer assembles a unique product from standardised parts (Sundbo, 1994). Other parts of the literature emphasise the interaction between the service provider and the customer, or 'the service encounter' (for a review, see Mattsson, 1994). In this case, it is more the customers' experience of the service relationship that is seen as 'flexible'. There is also literature that stresses the role of ICT in service innovation, including the important contribution on 'the reversed product life cycle' (Barras, 1986). Service innovation can be seen as a process innovation that leads to product innovation, usually in the form of flexible, adaptable services. These various theories are not altogether mutually consistent. Viewed in a broader societal context, however, they do add up to the idea of a possible choice in service production between flexible production and standardised production.

The productivity problem of in-person services

It is well known from the literature that there is a productivity problem connected with in-person services. Compared with manufacturing and agriculture, in-person services cannot easily improve productivity by means of mass production or an assembly-line system (cf. Gershuny, 1983; and for a discussion see Illeris, 1996). While industry and agriculture have become capital-intensive or characterised by economies of scale, in-person services seem to be necessarily labour-intensive. They require a minimum of available people, time and space. On top of this, as we have noted, these services are difficult to describe, to delimit and to divide. Their form and content are intertwined with professional judgement and, in the case of public services, with political thinking. The personal face-to-face contact between the service provider and the consumer necessarily contributes to the productive relationship. It is difficult to establish precisely what is produced and, for the consumer, what is acquired. For these reasons, it is difficult to set up precise criteria of productivity. Generally, therefore, it is difficult to see what a prime economic rationale behind standardised production of in-person services should look like, and flexibility remains a strong candidate.

The substitution effect

Productivity gains as a result of mechanisation and competition may be difficult to achieve in in-person services, although some in-person services can be replaced by manufactured products and the informal household production of services (cf. Gershuny, 1983). The substitution of manufactured products and household technologies for services may be as effective as mass-production methods for making services less expensive and increasing productivity. The substitution effect occurs in many in-person services, including services for the elderly: here, the clients generally have access to an increasing range of modern technology and can therefore participate to a greater extent in the flexible production of a range of services together with public and private service providers. For example, in services for the elderly, the use of email and Internet, personal alarm systems, satellite alert systems, mobile telephones etc. make it possible for the elderly to participate in the service production. In addition, a series of low technologies – in the bathroom for example – could help make the elderly more independent (see Cullen and Moran, 1992). Special toilets, cars, washing machines, vacuum cleaners, and telephones have already been introduced, making the elderly more self-dependent. In the future, robot cleaning could replace the manual kind. All this to some extent facilitates the informal household production of certain services, at least for the younger senior citizens. Further, like other social groups, some of the elderly now have greater economic, intellectual and physical resources as a result of higher incomes, less physically demanding working conditions during their active lives, more education etc. Potentially, therefore, the elderly become more independent and self-determinant, although they may also become more isolated. On balance, the substitution of products for services generally potentially tends towards more client involvement and thus to more flexible production, than towards mass production of services.

Fragmentation of services

In some areas in-person services have been characterised by fragmentation on the output as well as the input side. In spite of the collective management system, the home help policy of 'as long as possible in your own home' has led to the individualisation of work in the home. Three other factors potentially promote the fragmentation of the service production (see also the previous sections). (1) The recruitment of home helpers from weaker labour-market groups, particularly young women with few social or personal resources. (2) The fact that many elderly people today have more resources that encourage participation in service production, and cannot therefore easily be subjected to semi-professional attitudes. (3) The impact of budget cuts and accountability problems in the welfare state, which reduce the autonomy of the professional groups. These forces for fragmentation generally make it difficult, if not impossible, to establish consensus on a predominant style of home help that could be mass-produced. In-person services for the elderly are therefore compelled to remain in some way adaptable and flexible.

Job and client satisfaction

Flexible production can provide strong motivation for both employee and client. For the employee, flexible production implies influencing the working conditions, even if the client's influence is also important. For the client, Swedish studies among others have shown that elderly people normally prefer some form of flexible specialisation to 'assembly line' production. Szebehely (1995), for example, found that the elderly generally want to be able to anticipate and affect the services.

While the productivity problem is a more general phenomenon, the other four arguments are historically specific. Together they add up to a potential 'strategic' choice in the in-person services for the elderly between flexibility on the one hand – as an answer to the need for organisational programming – which seems to be less costly and more satisfactory, and mass-production techniques on the other, which are of course attractive to some firms that have spotted a market to be exploited here.

Conclusions

This chapter has attempted to uncover three production systems in welfare services, that is to say in the home help services in Denmark: a societal, interactive and an organisational system. The societal system was based on the social role of the housewife. The interactive system was based on the semi-professional role of the home helper. And the organisational system was based in strategic reflexivity and the role of the social entrepreneur.

The organisational system becomes more important as home help services are exposed to external demands. Home help services have to go through a process of organisational programming in order to meet these demands. Organisational programming involves strategic reflexivity, where the organisation reflects upon its role in the division of labour taking human and other resources into consideration. The social entrepreneur is a new management role that embodies strategic reflexivity at the middle-management level. She stimulates employees to reflect upon their role in relation to external demands and the organisational and strategic program-ming of them. Indeed, we have seen that there are many barriers to this role and we do not have any basis for concluding whether or not this role, as described through a case study, will be able to institutionalise.

11 Interventions, interdependence of stakeholders and implementation effectiveness

An empirical analysis of structuration processes in a multi-actor model

Marius T.H. Meeus and Leon A.G. Oerlemans

Abstract

This study concentrates on the problems of stakeholders (managers and workers) interacting in the context of implementing information technologies. The impacts of the interaction between structural and cognitive aspects of interdependence on implementation effectiveness in the context of process innovations are investigated. Structural interdependence means that managers have to rely upon workers' knowledge and skills to assimilate innovations which are institutionalised in worker participation and mostly represented by user participation in innovations processes. Cognitive interdependence means that (1) stakeholders monitor each others' interventions and judge their outcomes, (2) stakeholders can disagree about their judgements of intervention outcomes. Implementation effectiveness refers to the consistency, acceptance and quality of targeted organisational members' use of a specific innovation.

The data were acquired from plant and IT managers, from functional managers and non-managerial employees in six separate organisations implementing major process innovations. Our findings partially supported our hypotheses: (1) they showed mixed main effects of user participation on implementation effectiveness, while judgements of intervention outcomes, and level of dissent also showed mixed effects on implementation effectiveness, (2) effects of user participation on implementation effectiveness were contingent on the judgements of intervention outcomes and level of dissent on intervention outcomes. The internalisation approach versus an activity based relational approach are discussed.

Introduction

The sweeping statements of the IT community including its practitioners and researchers include: 'Information technology (IT) strongly impacts on modern economies. IT is supposed to determine the competitive edge of industrial and service sectors, and this awareness grows in tandem with the investments in IT'. Despite the doubts raised with such exaggerated self-promotion, the strategic importance of IT is undisputed and progress has been made in understanding it. However, implementing IT effectively remains problematic. Behind the almost infinite announcements of new releases of existing software, and new software, there is an enormous number of adaptation and learning problems to be tackled by adopting firms within a relatively short timeframe (Tyre and Orlikowski, 1997).

Increasingly, organisational analysts identify implementation failure, besides innovation failure, as the cause of many organisations' inability to achieve the intended benefits of the innovations they adopt. This applies to technology management in general, as a recent in-depth study revealed (Scott, 1997). The issue of organisational learning about technology and the development of technology core competence turned out to have a high priority among experts who had to identify the major unresolved management problems faced by companies. A broad range of innovations like quality circles, total quality management, statistical process control and computerised technologies often yield little or no benefits to adopting organisations, not because the innovations are ineffective, analysts suggest, but because their implementation is unsuccessful (Fleck, 1984; Lyytinen and Hirschheim, 1987; Majchrzac,1988; Cooper and Zmud, 1990; Zammuto and O'Connor, 1992; Hattrup and Kozlowski, 1993; Klein and Ralls, 1995; Klein and Sorra, 1996.

In this chapter we concentrate on the explanation of implementation outcomes, whether failures or successes. Klein and Sorra (1996) define innovation implementation as the process of gaining targeted employees' appropriate and committed use of an innovation. Implementation effectiveness refers to the consistency, acceptance and quality of targeted organisational members' use of a specific innovation. There are two issues explaining implementation effectiveness explored in this chapter. On the one hand, the impacts of user participation in innovation implementation will be discussed from a viewpoint of control and interdependence. On the other hand, we want to add to this rather common explanation of implementation outcomes, the social processes moderating impacts of user participation on implementation outcomes from a cognitive view on interdependence. The problem of user participation has been widely investigated, yet the cognitive processes involved are mostly disregarded. Extending the structural and motivational view of participation by employing a cognitive view of social dynamics induced by implementation processes yields a new view on change processes. Both aspects of organisational dynamics shall be discussed from a processual and agency view on power (Hardy and Clegg, 1996), stressing different but related dimensions of the *interdependence* of stakeholders during implementation processes. *User participation* is the outcome of a long historical process of shifting power balances between employers and employees and as such stresses the *structural dimension of interdependence* of both stakeholders. The *emergent awareness of innovation impacts* during the implementation process signifies the *cognitive dimension of interdependence* and emphasises that stakeholders actively monitor and judge the impacts of the implemented innovation, the underlying interventions and by implication the behaviours of stakeholders. Both dimensions of interdependence are supposedly sources of compliance, or commitment of stakeholders to organisational change, and its impacts. Therefore, our research question is: to what extent do structural and cognitive interdependence have an impact on implementation effectiveness?

The chapter is structured as follows. In our theoretical section we first explicate a number of explanations for implementation outcomes, and, second, assumptions for a processual and agency perspective on implementation are described. Subsequently, the research design is considered. Next, our results will be presented. We conclude the chapter with a summary and discussion of our major findings.

Theoretical framework

Theoretical background

Interdependence: definition and deficiency

The definition of interdependence originally concentrated on structural aspects of organising. Jobs or tasks are considered interdependent when the people performing them must rely on or collaborate with others to complete their work (Crozier, 1964; Lawrence and Lorsch, 1967; Thompson, 1956, 1976; Dean and Snell, 1991). The discussion about interdependence originated in the recognition that somehow ecological arrangements of technology, and work directly affect interaction patterns, visibility, and therefore at least how much and in what ways people are controlled (Hage, 1980: 402).

Two deficiencies of the interdependence debate are relevant for our research. Although interdependence mostly has a structural connotation, its effects on organisation design also imply power relations and dynamics between organisational units and the actors involved. Especially within the behavioural theory of organisations, where coalitions do not have a generally shared, consistent set of goals, the management of interdependence between these coalitions is crucial (Cyert and March, 1963).

Furthermore, the discussion concerning interdependence and change management developed separately, whereas they seem so strongly intertwined. The interdependence between stakeholders is often temporarily amplified and tested under the pressure of environmental change. Another flaw in the discussion on interdependence is the neglect of the cognitive dimension of interdependence. With the cognitive dimension of interdependence, we mean that stakeholders in change processes continuously monitor each others' behaviour and actions, related to the content and goals of the change process.

We contend that in the adoption and implementation of major new technologies, and the related management of change, a higher order task interdependence of management and workers plays a key role, in which structural and cognitive aspects interact and impact on the effectiveness of implementation processes. First, by the fact that during change processes impacting on organisational structures, the interdependence between management and workers is often supported by worker or user participation. Second, due to the fact that adaptive activities are often implemented in a short timespan (Tyre and Orlikowski, 1997) and impact heavily on the organisational microstructures, this induces actors to monitor and judge each others' actions more intensively than normal. Therefore outcomes of change management rely heavily upon the co-operation between management and workers. If workers do not use new technologies appropriately, corporate investments are detrimental for firm performance. Therefore the management of interdependence is considered as a key factor for the successful implementation of any innovation.

Interdependence of stakeholders and change activities

The issue of interdependence is inseparable from change *activities*, or interventions executed in the implementation stage. We focus on an activity based approach of interdependence because this prevents us from analysing organisational reality in terms of concepts that are miles apart from what actors in organisations do (Cooper and Kleinschmidt, 1986: 73). Dougherty (1996: 425) also suggests such a change in perspective on innovation in organisation theory. Instead of applying broad industrial organisation concepts like product or process innovation, radical or incremental innovation, *theories of organisation focusing on the innovation process should reflect the activities that are being organised, because at that level the managerial issues of difficulties with innovation occur.* However, Dougherty's advice is apparently opposite to that of another tendency in innovation studies, preferring encompassing concepts like organisational culture, innovation climate, and innovation-value fit (Zammuto and O'Connor, 1992, Romm *et al.*, 1991; Klein and Sorra, 1996). The main reason is that implementation research did not produce very consistent results, which caused an enormous variety in explanations of implementation outcomes (Cooper and Zmud, 1990; Alavi and Joachimsthaler, 1992; Klein and Sorra, 1996). Yet, in the formulation of hypotheses and additional research topics, both Klein and Sorra (1996: 1075) and Zammuto and O'Connor (1991: 718) return implicitly to an activity based approach, when they stress the importance of implementation strategies. After all, innovation climates and innovation value-fit cannot be achieved without actions and interaction of stakeholders.

From control to interdependence and dialectic of control

Change activities executed in the context of an innovation radically transforming organisational micro-structures initiate a specific dialectic between
managers and workers. Although the importance of managerial attitude
towards innovation and change cannot be overestimated (Hage, 1980;
Klein and Sorra, 1996), in the implementation stage their role is severely
restricted to resource allocation. At this stage users become the key actors,
testing the appropriateness, the usability, and task support of an implemented innovation. This specific aspect of the implementation process
explains the importance of our idea of interdependence of stakeholders,
managers as well as users. It is very important to stress the knowledge-
ability of users in this specific innovation stage. Managers who know little
of the technical and organisational ramifications of an innovation are likely
to delegate implementation management to subordinates who are more
knowledgeable. Hage (1998) suggests that in the implementation stage the
employees have the tacit knowledge needed to adapt what are general
technologies to the specifics of the particular production process and solve
the many problems that emerge in the implementation process. Consulting
workers not only overcomes resistance but, more critically, accomplishes
a more successful and quicker implementation.

These specific dynamics seriously restrict the amount of control to be
practised by managers over workers. This implies that in the implementation stage a hierarchical conceptualisation of power – managers deciding to
adopt an innovation and allocating the resources for that goal – loses its
explanatory value and needs elaboration. First, because there are important
restrictions to the managerial ability to (re-) allocate resources in the context of innovation projects: (1) most managers are restricted in time and
capability to absorb expert knowledge about specific innovations, and (2)
they cannot easily redistribute knowledge between employees concerning
the use of innovations. Second, because innovation processes are knowledge
intensive par excellence, managers have to rely upon knowledgeable implementation managers and users. This defines the first constituent of interdependence of managers and workers in the context of the implementation
process, implying that workers share managers' power to initiate change,
and that managers share workers' capabilities for the appropriate use of the
innovation. This issue has been back on the agenda of innovation research
recently (Dougherty and Hardy, 1996; Klein and Sorra, 1996). Dougherty
and Hardy (1996: 1149) contend that, 'the failure to acknowledge the role
of power, both in theory and in practice, may be one big reason why
researchers and practitioners do not understand the project-to-organisation
problems with innovation well enough to do much about them'. They crit-
icise (1996: 1147) discussions of power in the innovation and management
literature emphasising the personal power of individual managers, controlling resources, as the surface of power dynamics. Therefore, they make a

plea for the analysis of power lying in processes, in the structuring of meaning, creation of momentum and the legitimisation of actions.

Giddens' concept 'dialectic of control' (1984), which stresses this two-way and relational character of process control and power, provides such an account. Power within social systems presumes regularised relations of autonomy and dependence between actors or collectivities in contexts of social interaction (Crozier, 1964; Crozier and Friedberg, 1977; Willener, 1967). It addresses the issue of how the less powerful manage resources in such a way as to exert control over the more powerful in established power relationships. On the one hand this relational concept of power stresses that change processes are multi-actor processes with at least two interdependent stakeholders: decision-makers and targeted employees (Klein and Sorra, 1996; Freeman, 1984; Rowley, 1997). The stakeholder theory specifies the differential interest individuals or groups have in processes of change. Stakeholders are all those claimants inside and outside the organisation who have a vested interest in the problem and its solution (Mason and Mitroff, 1981: 43). Literature on implementation failure of information systems suggests that the key reason for implementation failure is the misunderstanding of the dynamics between stakeholders. The differences between changes in the user and organisational domains, and changes in the technical domain are underestimated. In other words, IT professionals have other ideas and beliefs about implementation than users have. This causes political 'games' between IT professionals and users avoiding using the information systems. These games are manifestations of interest-seeking strategies followed by certain stakeholders (Lyytinen and Hirschheim, 1987). Recently this view has been revived in the social constructionist approach of technical change, in which the social shaping of technology, especially the political nature of implementation processes have been put on the research agenda again (Van de Ven and Rogers, 1988; Badham, 1993; Williams and Edge, 1996). This view implies that successful change strategies ought to align the interests of different stakeholders. On the other hand, most of the aforementioned innovation researchers refrain from specifying the processes explaining the emergence of expectations, interests, commitment to the implemented changes, or conflict. Again a structurationalist notion of agency is useful here. Agency refers to the reflexive monitoring of action, which is a chronic feature of everyday action and involves the conduct not just of the individual but also of others. That is to say, actors not only monitor continuously the flow of their activities and expect others to do the same for their own; they also monitor aspects, social and physical, of the contexts in which they move (Giddens, 1984: 5). Now, one can easily imagine that in the context of change processes, this mutual monitoring of activities is intensified. This specific aspect of agency defines interdependence between stakeholders in cognitive terms and can account for the emergence of interest, expectations or conflicts.

Two types of interdependence and some hypotheses

Now we have summarised some backgrounds of the debate on interdependence as well as the flaws in conceptualising interdependence and in its linkage with the issue of change management, we can further develop our notions of structural and cognitive interdependence and our hypotheses.

Structural interdependence and implementation effectiveness

The first constituent of interdependence between employers and employees – institutionalised in worker or user participation – is directly relevant for preventing the dynamics of power hampering the implementation processes. User participation can be defined in terms of the activities users perform during system development and implementation (Mumford, 1986; Barki and Hartwick, 1994). User participation aims at the alignment of stakeholder interests in order to prevent intraorganisational conflict. User participation enables the expression of and negotiation about stakeholders' interests, and is supposed to be favourable to stakeholders' commitment to the implemented innovation, the ultimate goal of the implementation process. Empirical research supports these ideas. Gill and Krieger (1991) suggest that the extent to which managers depend on their workers for attaining their goals was the most important factor in determining the occurrence of user participation in IT-projects.

User participation is often considered as the best means to gain users' commitment. There has been an enormous amount of attention paid to the customer active paradigm in the general innovation practice and literature (Von Hippel, 1976; Teubal, 1976, 1998; Slaughter, 1998) and the long socio-technical (Mumford and Weir 1979, 1986; Guimaraes and McKeen, 1993; Gattiker and Larwood, 1995) and human relations tradition (Badham, 1993) advocating user participation in automation projects. Despite this, the scientific empirical findings are still rather fragmented and inconclusive about the impacts of user participation on implementation outcomes (Scott, 1997; Warmerdam and Riesewijk, 1988; Nutt, 1992; Guimaraes and McKeen, 1993; Wijnen and Van Oostrum, 1993; Barki and Hartwick, 1994; Fitzgerald, 1998).

Scott (1997) reports that customer involvement is considered one of the most complex technology management issues. Warmerdam and Riesewijk (1988: 91), Nutt (1992: 165) and Guimaraes and McKeen (1993) report positive relations between user participation and project outcomes, while Wijnen and Van Oostrum (1993) report a threshold effect. If working time spent on participation exceeded 30 per cent the effects became negative, whereas the number of participative activities had no significant relation with project outcomes.

One explanation for this might be that user participation is only meant as legitimisation for managerial change activities, and does not aim at a 'real' contribution of users. Changes in industrial relations law in the 1960s and 1970s, resulting in more democracy on the shop floor, obliged company management to let workers participate in strategic and organisational matters considered important. Another interpretation of the restricted impacts of user participation, is that it always suffers from the 'fallacy' of misplaced concreteness, which is counteracted during the implementation process. Hage (1980: 220–6) states that the full impact of innovations on social structure is felt only during the implementation stage in contrast to the initiation or development stage. All the tensions and conflicts inherent in social change appear. Whereas visions and high ideals characterise the initiation stage, the implementation stage is characterised by the opposite – dirty politics and disillusionment. During implementation, the expected changes become reality, causing shifts in the distribution of roles, functions, tasks, and status. Those changes in the microstructure often do not meet the expectations of stakeholders, causing an unwillingness to co-operate with change and disagreements about the change process.

Despite the mixed effects, the practice of user participation in IT projects is very widespread. Daniel (1987: 282) reported that user participation occurred according to 40 per cent of the (N = 2019) interviewed managers in the UK. Warmerdam and Riesewijk (1988) reported, in a non-representative sample, that in 73 per cent of the projects (N=67) there was user participation. Gill and Krieger (1991) investigated the diffusion of user participation in twelve European Union member states. In 75 per cent of the Dutch interviews user participation took place. Yet, we think that this diffusion of user participation mainly reflects the development of industrial relations law stressing the role of worker councils and their right to influence important strategic decisions affecting workers' job certainty (Gill and Krieger, 1991).

Cognitive interdependence and implementation effectiveness

The second, mostly disregarded, constituent of the interdependence of stakeholders relates to the emergence of interests (which obviates the idea of the neutral actor, and transforms actors into stakeholders) and the satisfaction of expectations. Sectional interests cannot emerge, expectations cannot be satisfied without stakeholders' continuous *monitoring and judgement of each other's actions and its outcomes*. Targeted employees in particular are given the opportunity to follow and influence the change process. They participate in the project team, in work groups, or perhaps even in the steering committee, or via worker councils. They know about the ins and outs of the project and communicate these, via periodicals to their colleagues. Of course, the extent to which outcomes of actions meet stakeholders' expectations determines the emergence of interest, of interest seeking strategies, and urges influencing of interest groups' actions. So, the achievement of their respective interests urges stakeholders to continuously monitor and judge each others' actions, which attach stakeholders' actions to each other.

Our combination of this *activity based* approach with the constituents of interdependence builds on findings of behavioural organisation theory which suggest that *motives follow action* (Cohen and March, 1974: Cohen *et al.*, 1972). The formation of motives is mediated by the cognitive process of stakeholders' judgement and monitoring of each other's *actions*. User participation enables stakeholders to institutionally monitor and judge each others' actions. In tandem, they engender the motives for compliance and/or commitment to innovations which determine implementation effectiveness (Klein and Sorra, 1996: 1056).

In contrast to approaches viewing the implementation process as primarily one of, more or less, passively applying pregiven technologies and adapting to their requirements, we emphasise the importance of the user-firm as a major source of technological innovation. Implementation is viewed as an active process of customising and developing working technological systems to fit organisationally defined requirements (Badham, 1993; Gerwin, 1988). In this view, the purchase of a particular information system or advanced manufacturing technology involves the acquisition by the firm of a technological resource, it does not become until several interventions have been initiated. The implementation of information systems in an organisation demands the execution of a combination of different interventions. These are defined as rational instances of social action of corporal beings in the ongoing process of events in the world (Giddens, 1976: 75). Interventions are sets of structured activities in which selected organisational entities are changed by selected organisational units (target groups or individuals) defining and implementing the activities (French and Bell, 1978). We consider every planned intervention as being a connection between an activity and an organisational entity (e.g. task, organisation charts, level of formalisation etc.) which the intervenor wants to change from one state into another (Meeus, 1994: 83).

The monitoring and judgement of interventions builds on two instances of cognitive interdependence: (1) targeted employees' judgements of the outcomes of interventions, and (2) the level of dissent between targeted employees' and intervenors' judgements of intervention outcomes. The first instance of cognitive interdependence reflects the notion that stakeholders judge intervention outcomes individually. If interventions have outcomes which do not meet stakeholders' expectations, the probability of negative judgements increases (Gerwin, 1988). This might cause a negative effect on the implementation effectiveness. Implementation effectiveness refers to the consistency, acceptance and quality of targeted organisational members' use of a specific innovation. Targeted organisational members (or targeted users) are individuals who are expected either to use the innovation directly (e.g. production workers, or office workers) or to support the innovations' use (IT professionals, production supervisors, management) (Klein and

Sorra, 1996). The second instance of cognitive interdependence takes into account that both stakeholders – targeted employees as well as intervenors – do not necessarily agree on the outcomes of interventions. The level of dissent on intervention outcomes is also supposed to moderate impacts of user participation on implementation effectiveness. The level of dissent indicates the extent in to which stakeholders agree on intervention outcomes. If managers judge outcomes more positively than targeted employees did, than intraorganisational conflict might arise, which might hamper implementation effectiveness. Therefore we hypothesise:

> *Hypothesis 1*: Implementation effectiveness is positively affected by (1) the level of user participation (structural interdependence), (2) positive judgements of outcomes of interventions, and (3) low dissent on intervention outcomes.

To specify the types of interventions whose outcomes are monitored and judged by stakeholders several tools for development and implementation of information systems and advanced manufacturing technology (CAD/CAM/CNC) were analysed. This revealed that three different types of interventions can be distinguished: process, technical and organisational interventions (Meeus, 1994). *Process interventions* concentrate on project management of the change process, for instance: the decisions about the choice of different types of interventions, the structuring and timing of interventions, as well as the allocation of financial means between hardware, software and training, and communication. Because process interventions define the project charter, they have a strategic function in the start-up phase and a support function during the implementation process. *Technical interventions* aim at an improvement of the technical IT infrastructure, and concern the implementation of hardware as well as software. In their review of implementation research Cooper and Zmud (1990) consider the lack of attention to IT technological characteristics as one of the major deficiencies. A second type of intervention, which was stressed in most studies, is *organisational interventions*. These interventions adapt the information technology to the local organisational context to meet the needs of the user organisation. Organisational interventions concern changes in the division of labour between departments, changes in the tasks and jobs, career opportunities of individual workers, and training. Initially, organisational interventions became more important because the trend towards more flexible and automated production systems, associated with a development towards a more organic, flexible structure turned out to hamper firm performance (Lawrence and Lorsch, 1967). Later, the importance of organisational interventions was stressed because of the deskilling caused by computerised technologies (Braverman, 1974; Hage, 1998). Alavi and Joachimsthaler (1992) estimate that user-situational variables – to be affected by the aforementioned interventions – can improve the implementation success rate by as much as 30 per cent.

The complexity of change processes is determined on the one hand by the involvement of more than one stakeholder, and on the other hand by the fact that there are multiple interventions impacting on the project outcomes. The simultaneous implementation of multiple interventions is caused by the fact that changes in one structural component of an organisation (strategy, technology, structure) necessitate adaptive changes in other components, otherwise the integration necessary for survival might be lost (Zucker, 1996: 167). Thus, positive outcomes of technical interventions might be counteracted by the neglect of organisational adaptations enhancing effective use of the new technology. Another possibility is that the intended goals of user participation are frustrated due to the fact that stakeholders were unable to negotiate effectively, because interventions were implemented very poorly, or because the management did not take the users who participated seriously. In such situations the reverse side of user participation might appear.

Barki and Hartwick report that users who participate in the development process were likely to develop beliefs that the new system is good, important and personally relevant. They also report a positive impact of user participation on levels of user involvement and attitude (1994: 75). Robey and his colleagues (1989, 1993) hypothesised and confirmed that individuals who participate in the systems development process will have greater influence on the information systems (IS) design decisions, be involved in more disagreements or conflicts (with other members of the development group), but be more likely to resolve these conflicts to their satisfaction. Barki and Hartwick (1993) tested these hypotheses and report paths linking user participation to both influence and conflict, as well as a path from user participation through influence to satisfactory conflict resolution.[1]

If outcomes of interventions are judged positively, we expect this to strengthen the commitment of users to the change process, which intensifies the impact of the user participation on the implementation effectiveness. Consequently, if outcomes of interventions are judged negatively, this weakens user commitment, which lessens the impact of user participation on implementation effectiveness. Therefore we hypothesise:

Hypothesis 2: The relationship between the level of user participation and implementation effectiveness is moderated by judgements of intervention outcomes.

The effects of dissent between stakeholders on the outcomes of interventions possibly have the same kind of moderating effect on the user participation–implementation effectiveness nexus. On the one hand, disagreement between stakeholders causes lower levels of commitment and therefore counteracts any potential positive effects of user participation on implementation effectiveness. Low dissent, on the other hand, causes higher levels of commitment and therefore reinforces positive

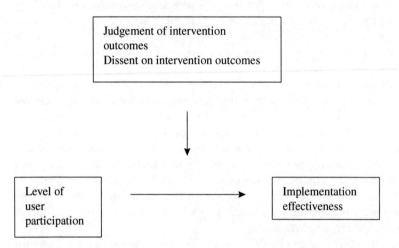

Figure 11.1 A conceptual model for an interventionist explanation of
implementation effectiveness

impacts of user participation on implementation effectiveness. Therefore
we hypothesise:

> *Hypothesis 3*: The relationship between the level of user participation and
> implementation effectiveness is moderated by the level of dissent on
> intervention outcomes between stakeholders (see Figure 11.1).

Method

Respondents, data acquisition, type of organisation and automation

In the period between 1991 and 1993 we gathered data on the innovation
processes in six organisations. Table 11.1 displays our research population.
The number of workers and managers and the type of automation in the
table indicate the number of respondents. The workers filled in a question-
naire, and the managers were interviewed by us using a structured ques-
tionnaire. The respondent workers were approached by the management,
and were selected according to two criteria, they were: (1) employed at a
department in which the automation was implemented, (2) involved in the
change process as a member of the work or project groups preparing the
project.

The investigated automation projects had quite different ramifications,
which need some clarification. In the municipalities, we investigated the
implementation of administrative software in the department of population
administration. This automation was enforced by Dutch law and aimed at

Table 11.1 Number of responding stakeholders, types of automation and types of organisations

Type of organisations	Type of automation	Number of responding stakeholders	
		Employees	Intervenors
Machine building (machine 1)	Enterprise resource planning	32	4
Machine building (machine 2)	Enterprise resource planning	46	3
Engineering firm	Computer aided drafting	67	3
Municipality 1	Automation population administration	13	3
Municipality 2	Automation population administration	25	3
Municipality 3	Automation population administration	10	3

a centralisation and harmonisation of the administration of the local population in all Dutch (ca. 689 at that time) municipalities. This administration is the basis for the production of identity cards and passports. It regulates the paying of local taxes, the right for certain payments e.g. for welfare benefits. This population administration also determines the size of the municipal budget. National government determines the municipal budgets by the number of inhabitants among other things. So this automation project was crucial for local authorities. The project impacted strongly on the tasks of civil servants. A separation between front and back office work emerged due to the separation of more complex and time consuming back office procedures and simple front office work e.g. the handing over of identity cards, which is basically data entry. Due to this project two departments (population affairs: data entry and handing over of e.g. birth certificates and identity cards), (civil affairs responsible for elections) merged. Another effect was that the new department became the core business of the municipalities, because they control the primary process and link the organisation to a number of users of their data e.g. retirement insurance companies, land registration, address brokers etc.

The implementation of an enterprise resource planning system (ERP) in two machine tool companies, both belonging to a large Dutch electronics producer, were investigated. The combination of the integration of activities between departments and a higher visibility of (dys-) functions of workers' activities proved to be essential. In both projects, the material resource planning was linked with the production and order planning on one hand, and with engineering and sales on the other. In both projects the level of communication and alignment of procedure underwent dramatic changes. ERP cross-cut the informal organisation, and new structures emerged causing a number of tensions and conflicts between departments, and with the management. Also the customisation of the software was a difficult and troublesome process. Despite the availability of customisation methods such as dynamic enterprise models, numerous parameters were not usable in default settings, and data conversion was often contingent on different methods of book keeping, or methods of cost calculation.

In an engineering firm the implementation of computer aided drafting (CAD) was analysed. In this company this meant a break away from the artisanal drafting tradition. The drawing boards and pencils were replaced with large computer screens, mouses and systems. The work space was completely reorganised, which impacted enormously on the informal organisation. Also, the quality of work was strongly affected, especially because of the new ways of handling or drafting rules, the storage and retrieval of data. This altered the old routines dramatically.

All automation projects we studied were significant departures from work practice and organisational routines. Therefore, a lot of ambiguous expectations must have emerged, which were only partially satisfied. Relations within departments and between departments changed greatly as well as the tasks and responsibilities of individual workers.

Measures

We take implementation effectiveness according to the targeted employees as our dependent variable. This is because: (1) targeted employees are the intended users, (2) if targeted employees would not confirm the quality of the new information system it will probably be underutilised, no matter what the management's judgement would be, (3) the judgement of implementation effectiveness by the management has no direct effects on the actual usage of information systems. This measure stresses the successful adoption of the change process, as well as the product (Bemelmans, 1991: 322). Successful adoption is considered to be a function of quality and acceptance of the process and the innovation (here the information system). The indicators were taken from De Brabander and Thiers (1983).

We use three different independent variables: (1) user participation as an indicator for structural interdependence, (2) the judgement of outcomes of three types of interventions, and (3) the dissent between stakeholders on three intervention outcomes. The second and third independent variables represent the two instances of cognitive interdependence. The judgement of *outcomes of technical interventions* was measured as the evaluation of changes in the technical features of production and/or administrative process, including hardware as well as software, caused by technical interference. Outcomes of technical interventions were measured in terms of usability of the computerised technologies: the efficiency increase due to the use of the system, and the user friendliness of the system. The *outcomes of process interventions* are defined as the changes caused by activities facilitating the change process as such. Process interventions concern the organisation of the project (Van Dijck, 1974; Wentink and Zanders, 1985); and communication (Salançik, 1977; De Brabander and Thiers, 1983; Warmerdam and Riesewijk, 1988). The outcomes of these interventions were measured in terms of items concerning operating procedures within the project, the formalisation of decision making and the knowledge of organisation, finance and goals of the project. The *outcomes of organisational interventions* are defined as changes in tasks, job complexity and labour conditions caused by organisation development (French and Bell; 1987; Wentink and Zanders, 1985; De Brabander and Thiers, 1983), training and education (Daniel, 1987; Chao and Kozlowski, 1986; Chaykowski and Slotsve, 1992).

The measures of dissent on implementation effectiveness, and dissent on the outcomes of interventions are based on subsets of items used above (see Annex). The mean score of the intervenors was determined for each item. Then the scores for targeted employees were recoded if the intervenors had a more positive assessment than the targeted employees did. The value 1 was assigned if there was dissent, the value 0 if there was consent. Subsequently, the compound scores were determined by adding up the scores for all items concerning implementation effectiveness, outcomes of technical, process and organisational interventions (see Annex).

Analyses

To test the hypotheses of this study, which predict that the relationship between user participation and implementation effectiveness is moderated by judgement of intervention outcomes, as well as dissent between stakeholders on intervention outcomes, we used hierarchical regression analysis (Tables 11.2–11.7). In relation to hypothesis 1, which tested first-order effects of user participation, judgement of intervention outcomes and dissent on intervention outcomes on implementation effectiveness, we entered user participation in the equation in the first step, the judgement of intervention outcomes in the second step and dissent on intervention outcomes was entered into the equation in the third step. For tests of moderation, in hypotheses 2 and 3, a similar procedure was applied (Cohen and Cohen, 1975; Miller and Droge, 1986; Stone and Hollenbeck, 1989). In step one, user participation was entered into the equation. In step two, and/or three we entered the particular moderator(s) in question. Hypothesis 2 was tested by first entering user participation into the equation, and the judgement of intervention outcomes was added. Finally, in step three, the cross-products of user participation with the moderator variables (e.g. user participation by judgement of outcomes of technical interventions) were entered as a set. Evidence of moderation exists when interaction terms account for a significant incremental variance in a dependent variable, either individually, signified by the values of the betas, or collectively, signified by the incremental F-statistics.

Table 11.2 Results of hierarchical regression analysis of user participation, judgement of intervention outcomes and dissent on intervention outcomes on implementation effectiveness.

Steps and variables	Machine 1			Machine 2			Engineering firm		
	M1	M2	M3	M1	M2	M3	M1	M2	M3
1 Level of user participation (UP)	−0.10	0.20	0.24	0.54*	0.56*	0.61*	0.27	−0.03	0.05
2 Judgement outcomes technical interventions		0.43*	0.55**		0.54*	0.43†		0.06	0.04
Judgement outcomes process interventions		0.40*	0.15		−0.25	−0.41		0.27	0.21
Judgement outcomes org. interventions		−0.17	−0.14		−0.08	0.09		0.33†	0.41*
3 Dissent outcomes technical interventions			−0.06			−0.31			0.04
Dissent outcomes process interventions			−0.45*			−0.15			−0.28†
Dissent outcomes org. interventions			0.25†			0.10			0.11
R^2	0.01	0.38	0.51	0.30	0.64	0.72	0.07	0.28	0.35
ΔR^2	0.01	0.37	0.13	0.30	0.34	0.08	0.07	0.21	0.07
F	0.33	5.72**	7.16**	7.10*	6.22**	4.11*	2.70	3.16*	2.26†
N		31			29			36	

Notes: †p <.10, *p<.05, **p<.01.

Results

Tables 11.2 to 11.7 present the results for: (1) the main effects of user participation, judgements of intervention outcomes, dissent on intervention outcomes on implementation effectiveness, (2) the effects of the interaction between user participation and judgement of intervention outcomes on implementation effectiveness, and (3) the effects of interaction between dissent on intervention outcomes and user participation on implementation effectiveness.

Hypothesis 1

Our results displayed in Tables 11.2 and 11.3 reveal that user participation had mixed effects on implementation outcomes in two of six cases. User participation either had significant negative effects (Table 11.2: machine 2),

Table 11.3 Results of hierarchical regression analysis of user participation, judgement of intervention outcomes and dissent on intervention outcomes on implementation effectiveness

Steps and variables	Municipality 1			Municipality 2			Municipality 3		
	M1	M2	M3	M1	M2	M3	M1	M2	M3
1 Level of user participation (UP)	0.35	−0.05	0.87*	0.26	0.35†	0.20	0.55†	0.34	0.30
2 Judgement outcomes technical interventions		0.69	0.04		−0.01	−0.09		0.44*	0.44*
Judgement outcomes process interventions		0.30	0.60*		0.15	0.09		−0.09	−0.25
Judgement outcomes org. interventions		−0.27	0.93*		0.40*	0.44*		0.65*	0.47*
3 Dissent outcomes technical interventions			−0.04			−0.26			−0.56†
Dissent outcomes process interventions			−0.44†			−0.47*			−0.17
Dissent outcomes org. interventions			1.30*			−0.15			0.66†
R^2	0.12	0.56	0.96	0.07	0.23	0.42	0.30	0.69	0.84
ΔR^2	0.12	0.44	0.40	0.07	0.16	0.19	0.30	0.39	0.15
F	1.23	1.91	10.76*	1.73	3.21†	4.99**	3.45†	7.64*	10.74**
N		10			25			9	

Notes: †$p < .10$, *$p < .05$, **$p < .01$.

or it had significant positive effects on implementation effectiveness (Table 11.3: municipality 1). However, user participation produced a significant F-value in two instances (Table 11.2: machine 2; Table 3: municipality 3).

Furthermore, our results revealed that higher scores on judgements of intervention outcomes, in case of significant betas, contribute positively to implementation effectiveness. In machines 1 and 2, respectively the engineering firm (Table 11.2) more positive judgements of technical interventions were conducive to implementation effectiveness. In municipality 1–3 (Table 11.3) the more positive judgements of organisational interventions in particular favoured implementation effectiveness. Adding the judgement of intervention outcomes to the model produced higher and significant F-values in five of six organisations, indicating that this operationalisation of cognitive interdependence has explanatory value for implementation effectiveness.

Table 11.4 Results of hierarchical regression analysis of interaction of user participation and judgements of intervention outcomes on implementation effectiveness

Steps and variables	Machine 1			Machine 2			Engineering firm		
	M1	M2	M3	M1	M2	M3	M1	M2	M3
1 Level of user (UP) participation	−0.10	0.05	1.14**	0.54*	−0.29	−0.42*	0.27	0.26†	0.31*
2 Judgement outcomes technical interventions		0.52**	0.31*		0.55*	0.49*		0.09	−1.02†
Judgement outcomes process interventions		0.40*	2.18**		−0.06	−0.08		0.20	0.42
Judgement outcomes org. interventions		−0.03	−0.03		−0.40*	−0.40*		0.35†	1.36*
3 UP* judgement outcomes technical interventions			−0.50			−0.18			1.33*
UP* judgement outcomes process interventions			−1.66**			0.17			−0.13
UP* judgement outcomes org. interventions			−0.00			−0.99			−1.38*
R^2	0.01	0.26	0.63	0.30	0.53	0.67	0.07	0.35	0.46
ΔR^2	0.01	0.25	0.37	0.30	0.23	0.14	0.07	0.28	0.11
F	0.33	4.95*	11.28**	7.10*	9.02**	10.44**	2.70	4.28**	3.55**
N		31			18			36	

Notes: †$p <.$, *$p<.05$, **$p<.01$.

With regard to the collective influence of dissent on outcomes of technical process and organisational interventions on implementation effectiveness hypothesis 1 is confirmed by all analysed projects. The explained variance increased significantly in each estimated model, and our analyses also generated significant F-values when dissent variables were added in the model. Inspection of the beta signs of the individual effects of dissent on intervention outcomes revealed that dissent on intervention outcomes significantly improves the explained variance in eight of sixteen cases. In five instances, the signs of the betas were negative – as expected – while in three instances the betas had positive signs. This implies that although workers disagreed with their superiors with regard to the outcomes of organisational interventions, this does not hamper implementation effectiveness. This phenomenon occurs especially in the case of organisational interventions (Table 11.2: machine 1; Table 11.3: municipality 1, 3).

Overall, these findings partially support the prediction of hypothesis 1. The most important finding is that the cognitive interdependence supersedes the structural interdependence in the explanation of implementation effectiveness. User participation contributes less to the explained variance of implementation effectiveness than judgements of intervention outcomes and dissent on intervention outcomes. Contrary to our expectations, however, user participation had a negative beta in one case (machine 2), and three of the eight significant betas of dissent variables had a positive sign. This suggests that negative impacts of user participation, and disagreement between managers and workers need not necessarily hamper implementation effectiveness.

Hypothesis 2

Tables 11.4 and 11.5 present the results for the effects of the interaction between user participation and judgements of intervention outcomes on implementation effectiveness. In all cases, the regression equations showed evidence of moderation. Entering the cross-products of user participation with judgement of outcomes of respectively technical, process and organisational interventions as a set, accounted for significant incremental variance in implementation effectiveness. In three cases (Table 11.4: machine 2; Table 11.5: municipality 1 and 3) the incremental variance of implementation effectiveness had a collective cause, signified by significant changes in F-values. In three cases (Table 11.4: machine 1 and engineering firm; Table 11.5: municipality 2) evidence of moderation was caused by individual interaction terms with significant betas accounting for significant incremental variance of implementation effectiveness. We found, contrary to our expectations, mixed effects of the interaction terms on implementation effectiveness. Three of four significant interaction terms had negative betas, suggesting: (1) that the conjuncture of non-participation and negative judgements of intervention results favoursimplementation effectiveness, and (2) that a conjuncture of positive judgements of intervention outcomes and user participation can hamper implementation effectiveness. Closer inspection of the raw data (cross-tables of participating versus non-participating users with levels of judgements of intervention outcomes) broadly confirmed this. In machine 1 negative judgements of interventions outcomes by participants and non-participants were in the majority, and user participation correlated negatively (Spearman Rho -0.44, p<.01) with the judgement of process interventions. In the engineering firm (Table 11.2), the non-participating users judged outcomes of organisational interventions systematically more negatively than the participating users (Spearman Rho 0.31, p<.01). In case of municipality 3 the same conjuncture of negative judgements of organisational interventions among non-participants and positive judgements among participants was found (Spearman Rho 0.18, n.s.).

Table 11.5 Results of hierarchical regression analysis of interaction of user participation and judgements of intervention outcomes on implementation effectiveness

Steps and variables	Municipality 1			Municipality 2			Municipality 3		
	M1	M2	M3	M1	M2	M3	M1	M2	M3
1 Level of user participation (UP)	0.23	−0.06	−0.27	−0.16	−0.09	0.04	0.55†	0.34	0.30
2 Judgement outcomes technical interventions		0.60†	0.66*		−0.13	−0.26		0.65*	0.44*
Judgement outcomes process interventions		0.46†	−0.07		0.17	−0.11		−0.09	−0.25
Judgement outcomes org. interventions		−0.79**	−0.81**		0.39†	1.59*		0.65*	0.47*
3 UP* judgement outcomes technical interventions			—			−0.05			0.93
UP* judgement outcomes process interventions			0.61			0.55			−0.15
UP* judgement outcomes org. interventions			—			1.43†			2.28
R²	0.05	0.67	0.70	0.03	0.21	0.49	0.30	0.69	0.84
ΔR²	0.05	0.62	0.03	0.03	0.18	0.28	0.30	0.39	0.15
F	0.61	4.13*	3.33†	0.57	1.29	2.16†	3.45†	7.64*	10.74**
N		13			23			9	

Notes: †p <.10, *p<.05, **p<.01, — variable excluded tolerance 0.000 limits reached.

These findings partially support the prediction of hypothesis 2 that judgements concerning intervention outcomes moderate the effects of user participation on implementation effectiveness. However, most moderating effects of judgements of intervention outcomes had a different consequence than that we expected. The mixed signs for the moderating effects of judgements of interventions on user participation suggest that positive judgements of intervention outcomes can either strengthen or weaken impacts of user participation on implementation effectiveness. Furthermore, even a conjuncture of non-participation with negative judgements of intervention outcomes is possibly associated with higher levels of implementation effectiveness.

Hypothesis 3

Tables 11.6 and 11.7 show the regression results for the interaction of user participation with dissent on intervention outcomes and implementation

Table 11.6 Results of hierarchical regression analysis of interaction of user participation and dissent intervention outcomes on implementation
effectiveness

Steps and variables	Machine 1			Machine 2			Engineering firm		
	M1	M2	M3	M1	M2	M3	M1	M2	M3
1 Level of user participation (UP)	−0.10	0.02	−1.09**	−0.26†	−0.20	0.42	0.26	0.24	0.14
2 Dissent on outcomes technical interventions		−0.46**	−1.59**		−0.25†	−0.29		−0.12	0.97†
Dissent on outcomes process interventions		−0.38*	−1.38**		−0.07	0.32		−0.16	−1.26**
Dissent on outcomes org. interventions		0.23	0.40		0.14	1.54*		0.03	1.06
3 UP* dissent on outcomes technical interventions			1.41**			0.07			−1.16†
UP* dissent on outcomes process interventions			1.41**			−0.55			1.36**
UP* dissent on outcomes org. interventions			−0.31			−1.53*			−0.90
R²	0.01	0.42	0.71	0.07	0.16	0.26	0.07	0.12	0.36
ΔR²	0.01	0.41	0.29	0.07	0.09	0.10	0.07	0.05	0.24
F	0.33	4.98**	8.50**	3.28†	2.01	1.89†	2.70	1.04	2.31*
N		31			45			36	

Notes: †p <.10, *p<.05, **p<.01.

effectiveness. In five of six cases the regression equations showed evidence
of moderation. In these five cases, the entering of the cross-products of user
participation with dissent on intervention outcomes of respectively technical, process and organisational interventions as a set, accounted for significant incremental variance in implementation effectiveness. In one case
(Table 11.7: municipality 2) the incremental variance of implementation
effectiveness had a collective cause, signified by notable changes in F-values. In four cases (Table 11.6: machine 1, 2 and engineering firm; Table 11.7:
municipality 3) evidence of moderation was caused by individual interaction
terms with significant betas accounting for significant incremental variance
of implementation effectiveness. We found, contrary to our expectations,
mixed effects of the interaction terms on implementation effectiveness.
Three of six significant interaction terms had the expected negative betas,
suggesting: (1) that the conjuncture of non-participation and low dissent
on intervention results favours implementation effectiveness, and (2) that a

Table 11.7 Results of hierarchical regression analysis of interaction of user participation and dissent intervention outcomes on implementation effectiveness

Steps and variables	Municipality 1			Municipality 2			Municipality 3		
	M1	M2	M3	M1	M2	M3	M1	M2	M3
1 Level of user participation (UP)	0.35	0.49	0.20	−0.16	−0.45*	−0.47*	0.55†	0.19	0.13
2 Dissent outcomes technical interventions		−0.01	−1.55		−0.53*	−0.20		−0.81**	−0.70**
Dissent outcomes process interventions		−0.58	−0.50		−0.21	−0.68		0.00	0.00
Dissent outcomes org. interventions		0.02	0.11		−0.28	−0.02		−0.29†	1.10
3 UP* dissent outcomes technical interventions			1.48			−0.56			−0.81
UP* dissent outcomes process interventions			—			0.78			−0.11
UP* dissent outcomes org. interventions			—			−0.29			−0.33*
R^2	0.12	0.45	0.53	0.03	0.54	0.56	0.30	0.84	0.92
ΔR^2	0.12	0.33	0.08	0.03	0.51	0.02	0.30	0.54	0.08
F	1.23	1.25	1.11	0.57	5.59**	2.94*	3.45†	17.95**	23.52**
N					23				9

Notes: †$p < .10$, *$p < .05$, **$p < .01$, — variable excluded, tolerance limit 0.00 reached.

conjuncture of high dissent on intervention outcomes and user participation can hamper implementation effectiveness. In the other three instances, the interaction terms had positive signs, suggesting that participants with higher dissent, assessed the implementation effectiveness more positively than non-participants with low dissent. Again, we re-analysed our raw data and found that in machine 1 there was indeed a conjuncture of low dissent with non-participation, and high dissent with participation. This pattern was found in the Engineering Firm too.

Overall, these findings provide only partial support for hypothesis 3. The interaction effects of user participation with the dissent variables on implementation effectiveness were, in three out of six significant betas positive, contrary to our expectation. This finding suggests that in some cases user participation produces a certain disagreement between workers and management without hampering implementation effectiveness.

Discussion

This study clearly shows that the relationship between user participation and implementation effectiveness is anything but automatic. Although an integral part of best practice in the implementation of computerised technologies, user participation does not guarantee implementation effectiveness. The low number of significant effects and the mixed beta signs of user participation on implementation effectiveness leads us to question implementation theories considering user participation as a panacea for the generation of workers' loyalty and commitment towards change processes. Although user participation is ubiquitous, it is performed in many ways although there is a general tendency explaining its present development. The basis of user participation in the legally imposed industrial democracy in many advanced economies, in tandem with the enormous complexity of modern information technology challenged managerial capabilities, and poured off managerial hegemony in decision making. During radical change processes managers must, on the one hand, rely on the local knowledge of the firm's employees to implement new information systems successfully, while, on the other hand, they must reconcile the control over the transformation process, with workers' claims to 'take part' in the firm's decisions. This explains, partially, why principals in charge of such high-risk, radical transformations as the implementation of computerised technology develop a formal, legislative and externalist view of user participation concentrating on the acquisition of the employees' local technical, organisational and economic knowledge, and the legal obligations to be satisfied. In such a view of user participation, the side effects of employees engaging in influence activities are completely disregarded and no proficiency to deal with workers' interests and competencies is developed. Giving workers an opportunity to engage in influence activities and soliciting their information generates expectations with regard to the types of actions, their timing in the change process and their outcomes. Here lies the main source of the dialectic of control invoked by user participation. In the context of change processes, workers' expectations tend to become managers' obligations. For if workers' expectations are not met, their co-operative efforts and commitment are inhibited by latent processes moderating the effects of user participation.

In particular, our findings show that the judgement of intervention outcomes, as well as dissent on intervention outcomes, collectively moderate the effects of user participation on implementation effectiveness. Consequently, one's attention is bent toward the cognitive dynamics inherent to user participation. The moderating effects of judgements of intervention outcomes on user participation deflect attention to the quality of the interventions implemented by managers and IT professionals responsible for the project management; whereas the moderating effects of dissent on intervention outcomes deflect attention from participation to the outcomes of the interaction between intervenors and workers in terms of levels of dissent. Our research design in tandem with the operationalisations of cognitive interdependence makes clear that an externalist view of user participation measured in terms of level of involvement without specification of the cognitive processes, disregards the fact that non-participating users also judge intervention outcomes, and that these workers also int᠎ ᠎ct with project ᠎anagers during daily routines. The point is that non-participating users can be involved as much as the participating users. The internalist view developed in this study integrates those dynamics originated by interventions during change processes.

Our findings indicate three things. First, since the moderation effects were prevalent across all investigated implementation projects – the advanced manufacturing projects as well as the automation of administrative processes in municipalities – future researchers should continue to examine a broad variety of automation types, in a broad variety of organisations. However, in the research design one ought to restrict this variation in such a way that there are at least pairs of comparable organisations doing comparable automation projects. Such a research design enhances the generalisability of findings over multiple types of organisations and automation. On the one hand, much previous work in implementation studies has leaned very heavily on implementation of one specific type of software e.g.: advanced manufacturing technologies (Fleck, 1984; Dean and Snell, 1991; Cooper and Zmud, 1990; Wall *et al.* 1990; Zammuto and O'Connor, 1992), decision support systems (Alavi and Joachimsthaler, 1992), mainframe software package (Cale and Eriksen, 1994), EDI (Premkamur *et al.* 1994; Hart and Saunders, 1997) etc. On the other hand, much previous research has a broad meta-analytical character and compares implementation processes across different types of automation and organisation (Daniel, 1987; Ives *et al.* 1983; Lai and Mahapatra, 1997). Both types of research have contributed considerably to the state of the art scientific knowledge on implementation. However, as stated earlier, user participation in automation projects as a practice is widely diffused and probably has very different impacts under varying conditions. Yet there are few research projects in which impacts of user participation on implementation effectiveness are compared across different contexts.

Second, the effects of cognitive interdependence on implementation effectiveness pertained to all three types of interventions distinguished. Overall regression results, counting significant betas with the expected signs, showed first that the judgement of intervention outcomes appeared to be better predictors than the dissent variables, and second that significant judgement variables pertain to other interventions than the significant dissent variables. The judgement of organisational interventions and judgement of technical interventions both had seven significant betas with the expected signs in twelve regression equations. Whereas effects of judgement of process interventions on implementation effectiveness was significant in two cases. With regard to the dissent variables the pattern was the exact opposite. Dissent on outcomes of process interventions had six significant betas with the expected signs in twelve regression equations, whereas dissent on technical and organisational interventions had two, respectively zero significant betas with the expected signs in twelve estimations of implementation effectiveness. These overall findings suggest that the impacts of process interventions are relatively more sensitive to the interaction between workers and managers. Whereas the effects of technical and organisational interventions do not seem to be affected by this interaction between stakeholders, and depend more on the individual workers' expectations, preferences and beliefs determining the nature of his judgements of intervention outcomes.

Third, the mixed signs of the significant beta coefficients revealed that judgements of intervention outcomes and dissent on outcomes of interventions both attenuate and enhance the relationship of user participation with implementation effectiveness. Our supposition that cognitive interdependence would play an important role in moderating that relationship is justified in the light of the results. Although the dialectic of control provides a conceptual base for this study and accounts for most of our findings, it is a very generic insight in interaction dynamics, which clearly does not tell the entire story of how contextual moderation affects user participation. In particular, the empirical work of Barki and Hartwick (1994) and Robey *et al.* (1989, 1993) enhances a further specification of the dialectic of control and makes some unexpected signs of beta coefficients interpretable. For instance, the negative signs of the interaction term user participation * dissent organisational interventions, and user participation * dissent on technical interventions (Table 11.6: machine 2, Table 11.7: municipality 3).

Participants as compared to non-participants are more likely to believe that the new system is good, important and personally relevant which explains high scores on implementation effectiveness. Robey *et al.* found that relative to non-participants, participating users can deploy more influence activities, but they can also get involved in more disagreements or conflicts. This accounts for relatively higher scores of participants on dissent variables. But participants also have the opportunity to resolve their conflicts, which is not likely in case of non-participants, which in its turn explains low scores on dissent variables. So engagement in influence activities possibly produces higher scores on implementation effectiveness, in tandem with higher scores of the interaction term of user participation and dissent on intervention results in case of disagreements, and to low scores for participants on the interaction terms in case of conflict resolution.

We argued for the impacts of both types of interdependence, in terms of the formation of commitment, loyalty and expectations, taking into account only one type of resource (influence via user participation), omitting other types of resources enabling workers and managers to mutually enforce compliance with their goals and interests. A number of other resources such as status, educational level, which indicate to what extent workers can utilise this influence during negotiations with the management, could have led to a further refinement of our predictions. For instance positive judgements of participants with a low status probably has less impact on implementation effectiveness, than positive judgements of participants with a high status. Future research could expand this theory to capture more of the complexity inherent in the interdependence–implementation effectiveness nexus.

In assessing the contribution of our study, some caution should be exercised, because of the small samples of key informants and the number of projects that were analysed. Of course, a small project cannot be made bigger, but it would be better if there was more systematic variation between large, medium-sized and small projects because of the different group dynamics involved, and the different number of hierarchical levels involved. Especially the enormous diffusion of ERP in large organisations provides an excellent opportunity to do so.

When relating our results to the dispute on preferred research directions in implementation research – Dougherty's preference for an activity-based approach versus Klein and Sorra's inclination to convert to a holistic climate-culture approach – we infer that a further development of an activity-based approach is necessary. In the construct 'innovation climate' activities, structures, norms and values are conflated at the cost of analytical clarity. As a consequence the specification of learning implications are crushed under the apparent clarity of such holistic concepts. In an activity-based approach the actors' interventions can be evaluated separately, enhancing a specification of learning problems.

Note

1   Of course we have to be cautious with all the empirical findings on impacts of user participation, for they apply to: different stages (especially design and development) and different types of information technology (decision support systems, groupware, advanced manufacturing technologies including MRP, CAD, CAM, CNC etc.). Yet these findings are important because they all stress that participation does impact on innovation outcomes, although its impacts are moderated by mediating variables.

Annex

Measure	Scales, number of items, Cronbach's α (if calculated)	Type of scale, calculation of score
Judgement of implementation effectiveness	Evaluation of items concerning (0.54): • the acceptance of the system • the success of the change process • the satisfaction with the participation Items were taken from Brabander and Thiers (1983), Davis (1993)	A Likert 5-point scale was used: (1) disagree strongly, (2) disagree, (3) neutral, (4) agree, (5) strongly agree. An interval variable was computed by adding up all scores, and dividing it by three. The range of the variable is between 0 and 3.
Level of user participation	Respondents were asked whether they participated in the project directly, e.g. as a member of the project team, work groups etc.	A dichotomous scale is used: workers participated directly or they did not.
Judgement of outcomes of technical interventions	Evaluation of scales concerning: A The efficiency impacts of automation (0.76): • this system gives me the opportunity of doing a better job • this system makes my work more comfortable • this system increases my efficiency B The user friendliness of the hard- and software (0.87): • the screen is clear and can be read well • computer prints are well readable and understandable • the software is user friendly • displays, monitors and gauges and the computerized hardware are user friendly Items were taken from M. Campion (1988).	A Likert 3-point scale was used for item set A: (1) this is not the case, (2) this is slightly the case, (3) this is strongly the case. For the items of scale B a 5-point Likert scale was used: (1) disagree strongly, (2) disagree, (3) neutral, (4) agree, (5) strongly agree. An interval variable was computed by adding up the scores for item set A and B. For item set A scores ≥ 6 were recoded 1, else 0. For item set B all scores ≥ 12 were recoded 1, else 0. Finally the compound value was counted by counting the scores for item set A and B. The range of the score is 0–2.

Measure	Scales, number of items, Cronbach's α (if calculated)	Type of scale, calculation of score
Judgement of outcomes of process interventions	Scales based on evaluation of items: 1 concerning the operating procedure within the project (two items, 0.66) 2 concerning the formalisation and transparancy of decision making (one item) 3 knowledge of financial and organisational aspects of the project (5 items, 0.94) 4 knowledge of structure of cooperation in terms of the people who cooperate, and timing of activities (5 items, 0.94) 5 knowledge of the goals of activities within the project (two items, 0.94) Items were taken from Warmerdam and Riesewijk (1988).	For scale 1 a 5-point Likert scale was applied: (1) disagree strongly, (2) disagree, (3) neutral, (4) agree, (5) strongly agree. For item 2 four answers were used (1) intransparant decision making, unclear formalisation, (2) informal decision making but clear division of tasks, (3) high formalisation of decision making, partially put into practice, (4) high formalisation put into use. For scales 3–5, 4-point Likert scales were used: (1) unknown, (2) somewhat known, (3) reasonably known, (4) well known. An interval variable was computed by recoding raw scale scores: scale 1 scores ≥ 8 = 1, scale 2 scores ≥ 3 = 1, scale 3 and 4 scores ≥ 15 = 1, scale 5 scores ≥ 6 = 1. Scores lower than the values mentioned were recoded as 0. Subsequently we added up the transformed scale scores which produced the compound score. The range of the scores is 0–5.
Judgement of outcomes of organisational interventions	Scales based on evaluation of items concerning: 1 changes in learning opportunities in the job (4 items, 0.80) 2 changes in mental work load (2 items, 0.79) 3 changes in social contacts in the job (2 items, 0.79) 4 changes in work autonomy (3 items, 0.79) 5 needs for additional training. Scales 1–4 are based on Wentink and Zanders (1985) and Daniel (1987). Item 5 is taken from Brabander and Thiers (1983)	For scales 1–4 a 3-point scale was used: (1) deteriorated, (2) unchanged, (3) improved. For scale 5 a 4-point scale was used: (1) need much more instruction, (2) need some additional instruction, (3) no additional instruction needed, (4) additional instruction would be a waste of time and money. The compound score was calculated in two steps. First the scale scores were recoded: scale 1 score ≥ 8 = 1, scale 2 score ≥ 4 = 1, scale 3 score ≥ 4 = 1, scale 4 score score ≥ 6 = 1, item 5 score ≥ 3 = 1. All other scores were assigned the value 0. Subsequently we added up the transformed scale scores which produced the compound score. The range of the scores is 0–5.

Measure	Scales, number of items, Cronbach's α (if calculated)	Type of scale, calculation of score
Dissent on implementation effectiveness	Items concerning: • acceptance of the system • success of the project • alignment of interest of workers and management	For every organisation separately the mean scores of the intervenors were determined per item. Then the scores for each targeted employee were recoded with value 1 if the intervenors had a more positive assessment than the worker of the intervention outcomes. If this was not the case, the dissent score would be 0. Finally we added up the score to calculate the compound dissent score. The range for this variable is 0–3.
Dissent on outcomes of technical interventions	Items concerning: • user friendliness of the system • improvement of productivity • improvement of efficiency	Same procedure. The range for this variable is 0–3.
Dissent on outcomes of process interventions	Items concerning: • system – organisation fit • formalisation of decision making • problems with timely delivery of the system • communication problems between: it-professional, users, and managers	Same procedure. The range for this variable is 0–4.
Dissent on outcomes of organisational interventions	Items concerning changes in: • changes in responsibilities • changes in work pace • changes in selection of work methods • changes in control • changes in required qualifications	Same procedure. The range for this variable is 0–4.

12 Strategic reflexivity in industrial service innovation

Managing inter-organisational conflict

Jan Mattsson

Abstract

A great deal of innovation activity and its success is contingent on effective collaboration between departments in organisations. This chapter reports on a case study of innovation. It investigates a project to upgrade existing industrial test facilities, and illustrates one of the essential points of this book, namely that the structure of the innovation process has changed from planning and hierarchy to strategic reflexivity in social interaction. By using a qualitative, interpretative method, the underlying quality dimension of the innovation process was elicited from participant actors. These were categorised into three groups in terms of centrality *vis-à-vis* the innovation process, namely into struggling, instrumental and concerned actors. We argue that by using qualitative approaches to unravel key inter- and intraorganisational innovation obstacles, management can become more aware of how different quality dimensions connect during innovation work and are then able to allocate resources according to a clear set of priorities.

The growing importance of industrial services innovation

Although the industrial services sector represents an increasingly important growth sector in the economy, very little research has probed into the secrets of how new industrial services are initiated, constructed and implemented. Industry leaders (85 per cent) within the EC-countries are even convinced that the share of service content in industrial output will further increase profitability and productivity (Commission of European Communities 1989). Underlying this is the argument that service input adds innovation and expertise to industry.

A number of attempts have been made to classify services in, or to, industrial firms (Gummesson 1991). However, the hallmark of the service sector is heterogeneity. Therefore, the activities which should be included in what has been termed producer services are still open to debate.

Normally, producer services are those intermediate services that are marketed to firms or institutions for incorporation in the production of a good or another service (Daniels 1990). Typical examples would be financial services, technical support and maintenance services and professional services such as IT-services.

Only a tiny fraction of published articles integrate service and new physical product development literatures (de Brentani, 1991). Therefore, there is also a lack of research focusing on how new services in industrial processes are developed in conjunction with product innovations and upgrading of production equipment (Easingwood 1986). This chapter will draw on the findings of recent innovation theory related to services (Sundbo 1998a), and explores the nature of strategic reflexivity in industrial service innovation. The sub-components of strategic reflexivity have been defined as interaction, roles (in innovation work), complementarity (between roles) and innovation management (see Chapter 1). We will show the inter-relations between these components by modelling innovation as a quality generating process.

This chapter reports on a case study of innovation in industry concerning the upgrade of existing certifying test equipment in marine engine laboratories (for exhaust particles and emission). We argue that internal development such as this R&D project can be analysed as to its quality structure, which contains the underlying quality dimensions to be overcome by innovation work. In this way, the structure of the innovation process can be mapped and assessed as to qualitative content. A better understanding of the nature of internal development services would place them at the core of industrial activities.

The purpose of the chapter is thus to demonstrate the emergent and social nature of innovation processes on the micro-level and the inter-organisational conflicts that shape the outcome. We will show the inherently social nature of innovation and the way in which strategic reflexivity is acted out as a systematic negotiation between actors of their roles. We will be able to depict the fragmented and often conflict-based character of interaction when actors attempt to find these roles (see Chapter 1 in this book by Sundbo and Fuglsang).

We will structure the chapter as follows: first the method and its theoretical underpinnings will be elaborated. Second, the case study entitled 'the struggle between two computer camps' is presented. Third, we discuss the theoretical implications of our approach to innovation and the nature of inter-organisational collaboration from an innovation perspective.

Method

Theoretical assumptions

Innovation research is in need of improved methods to fully explore the complexity of the innovation process (Johne and Storey 1998; Edvardsson and Gustafsson 1999). Some studies have used the critical incident approach (Flanagan 1954) to collect a wider spectrum of empirical experience data (Stauss and Hentschel 1991; Mattsson 2000). Much service research, however, has used traditional sampling and survey techniques and only measures the output of the innovation process (Easingwood 1986). What is also needed, we argue, are methods that capture process aspects of innovation as they are perceived by participant actors in relevant settings (Gustafsson *et al.* 1999).

Consumer services are often contextualised by what the literature has defined as the service encounter (Bitner *et al.* 1990; Grönroos 1990). In industrial settings, however, taking the encounter as the focal point of interest will be somewhat misleading as activities occur in a complex process between different employees and external actors in a co-ordinated way. We will instead use the term innovation process (see Sundbo, Chapter 3, for a discussion on organisational principles of innovation) to describe our object of study. Each participant in this process, i.e., employees, managers and/or external actors has a different perspective (Mattsson and Benson-Rea 1993). Our aim is to elicit verbal material from highly involved participants (henceforth called actors) as a way to elicit key quality elements from innovation work (Edvardsson *et al.* 1995). We take subjectivity for granted and the efforts are concentrated on abstracting and interpreting quality elements that are mentioned by actors.

Data collection

Data was collected by first developing a case-specific phase model of the innovation process (i.e. not a theoretical model of the innovation process such as those suggested by Edvardsson and Mattsson 1992; Edvardsson 1997). The few words describing the different phases of innovation were then used as trigger words to generate verbal material from actors. Actors were selected by having each actor name those who were considered most involved. A saturation level of ten actors was reached. Each one of these actors was approached and asked to give a taped account of what had happened during the different phases of innovation. The verbal materials can thus be considered as unrestricted. When the respondent had completed his narrative, the tape was played back to him for additions or corrections. These were also taped.

After a preliminary analysis of the verbal material, another session was scheduled. Now, the researcher's interpretation was discussed and the respondent was given a second chance to moderate outcomes. The purpose was to 'ground' in an intersubjective way the interpretations made (Glaser and Strauss 1967).

Building a model of the innovation process

First, the accounts of respondents were compared as to content by classifying their verbal material in the different phases of the trigger word model. Depending on how extensively the material covered the phases of the model, respondents were categorised as more or less central. Actors with a similar degree of centrality were, for the purposes of analysis, grouped together depending on their role in the process. Three groups of centrality emerged; *struggling actors* being most central, *instrumental actors* as second most central and, finally, *concerned actors* as least central. The case analysis follows this order of centrality.

Qualitative cues are embedded in verbal text. They are expressed as value terms, attitudes, quality criteria and the like. The essence of the data analysis and interpretation has been to arrive at a joint terminology of focal quality expressions between the respondent and the researcher. By first using an unrestricted approach to generate verbal materials, and, second, by attempting to find a consensus terminology for key qualitative aspects in the innovation process, we can claim a reasonably robust way of eliciting quality elements. They will now be the topic in the following case presentation.

Case: the struggle between two computer camps

The case consists of the feasibility studies, decision process and installation work of a PC-based electronic control system for the automatisation and certification of test facilities for new marine engine prototypes. Testing of performance and quality standards (e.g. ISO standards) has become mandatory for major European industrial firms in face of the global competition. The duration of the case description is from January one year until late May the next year when a first facility was completed. During this period a struggle between two computer camps was going on, one camp advocating the upgrading of the existing minicomputer system, the other camp pressing for a decentralised PC-based system.

The actors of the development process have been categorised as to their centrality in three groups; 'struggling actors', 'instrumental actors' and 'concerned actors' (see Table 12.1). The group of struggling actors includes the responsible system expert (STDA) of the existing system and his manager in charge of technical computer systems (STD; both belonging to the mini computer camp) and the operating manager of the test facilities (STP; PC camp), and the manager of the entire support department (ST; to which all actors of this group report). The latter has been caught in the middle of the struggle.

The instrumental group includes two actors; the supplier's regional representative (L) and the internal test facility user (customer), the director of the marine engine profit centre (M). They have both been instrumental in the

Table 12.1 Centrality of actors in the innovation process and quality elements

Centrality of actor group	Quality element
Struggling actors	
STD, STDA	Rational computer development.
STP	Decentralised and user-friendly PC-system
ST	Good work climate
	Good decision
Instrumental actors	
L	Responsive and client-adapted software functioning
M	Certifying facility
Concerned actors	
Q	New environment and quality standards
A	New functional demands (on buildings)
STE	Correct economic reports
SI	Purchasing expertise

sense that they have pushed for the PC alternative. The third group of 'concerned' actors contains two staff managers for buildings and operational security (A) and quality (Q), one member of the purchasing department (SI) and the controller assigned to the support department (STE; reporting to ST). We will now illustrate strategic reflexivity during the innovation process.

Struggling actors

The manager in charge of technical computer systems (STD) gives the background.

> Technical development in the computer field is very rapid ... therefore we have to safeguard the investments in systems we already have (called phase 1 and 2) in an optimal way ... We therefore put forward a long term strategy for computer system development. The first step was the upgrading of the existing system that we termed phase 3.

The responsible system expert (STDA) adds: 'We wanted to lift the system step by step ... we introduced a modular way of thinking ... we worked according to plan. Our system fulfilled all technical requirements.' In other words, both actors propose the element of *rational computer development* as their main argument. They have a plan which is to be followed.

In contrast, the operating manager of the test facilities (STP) puts his case thus: 'There is a break-through going on in favour of PC-computers for applications that were earlier dominated by mini computers ... you see no end to it! Conflicts between traditional computer experts and new PC users arise in firms ... jobs are at stake ... complete programming departments will be closed down when the PC world is entered.'

He goes on to argue his case for a decentralised PC system.

> We got hold of a new PC-based software control system developed for the process industry (i.e., not industrial batch production) with very simple (image) programming tools. We saw the advantages of letting our system users learn this. After ten days, we had already installed a simple software test package . . . we wanted to show how easy the system was to operate . . . it was a great success . . . everybody was very impressed by the interface graphics.

These verbal materials illustrate the operating manager's quality elements; *a decentralised and user-friendly PC-system*. This contrasts sharply with the rational view advocated by the system expert (STDA): 'The role of the expert is needed . . . there are no maintenance-free computer systems.' These excerpts clearly demonstrate the clash between roles and role expectations.

The manager of the entire support department (ST) tried to find a common ground for the two camps. He recalls an all-night meeting the day before the final decision was to be taken.

> We tried to put together a neutral document that we all could agree on. They were hitting the roof one after the other. I tried to put the facts together . . . the pros and cons of both alternatives. Nobody was satisfied . . . and everybody was very tense . . . we tried to reach as far as possible with a technical issue. I had strong feelings of anxiety beforehand, as I inherited this conflict when taking up this new job . . . the deeper you penetrate an issue, the more you see that there are never any certain facts.

The experience depicts the boundaries of rationality, as all 'facts' have to be derived from presuppositions that are based on underlying individual values and attitudes, which are not directly translatable. In this sense, the department manager takes on an impossible task to secure a united department position on computers and, above all; *a good work climate*. Here management attempts to 'direct' the personnel by finding new roles that the entire department can agree with. The social tensions are obvious, however.

Finally, he is forced to take a stand. He recapitulates:

> It was very difficult to take a clear stand on facts alone . . . the PC system included standard, and much cheaper, components . . . no need for expensive service support contracts with suppliers. Our systems become less vulnerable to breakdowns . . . we can easily move them between different facilities . . . the PC alternative does not have the same technical capabilities as the centralised mini computer system

... but we have not been able to show that we need this computing capability.

So, in the end, he favours the PC alternative and he arrives at *a good decision*. Forecasting is less useful and vision and foresight decide the outcome. Innovation management has indeed displayed strategic reflexivity. The circumstances that necessitated this decision are well illustrated by the instrumental actors.

Instrumental actors

Two actors have been instrumental in implementing the PC based system: the PC system supplier (L) and the director of the profit centre (M). The regional representative (L) tells us how he came into the picture.

> Our English supplier organised a road-show of industrial process control systems that we normally are not able to demonstrate. The manager of the test facilities (STP) discussed certain control equipment and not control data collection (systems) ... but most of the signals that were to be controlled also had to be logged ... so why have two separate systems taking in the same signals? We started to look at how we could collect and log all signals with our concept ... that was the real starting point.

The representative (L) continues to discuss the prototype installation of the software package. 'After the demonstration nobody had any doubts about the control aspect of our system ... the remaining question mark concerned how our concept would be able to log all control data. We quickly made a software application in EXCEL ... and we were able to show that our system could generate the same reports as the existing system.' This verbal material signals *a responsive, client adapted software* as the core quality element to secure the order. The innovation is initiated by an ad hoc social interaction and by demonstrating a prototype installation its manager becomes convinced of its practicality. Knowledge and flexibility drives innovation (see Chapter 1 in this book)

As a user of the test facilities, the director of the profit centre (M) had from the first day on his new job in January taken a keen interest in the project. He recalls: 'I discussed the functioning of the existing system with my operators ... they were very dissatisfied ... the system had regular breakdowns ... they did not find it necessary to expand the computer part of the system ... rather, they much preferred an upgrading of the measurement equipment.' During the important decision-meeting the director presses the issue of the advantages of a PC-based system. 'So, we decided to scrap the old system completely ... there were two decisive reasons for this ... flexibility and operating costs. We had to install a functioning certifying facility to make effective use of it . . . we simply wanted another control system . . . the customer is more important than the expert!' In other words: the installation of *a functioning certifying facility* is of the utmost importance for meeting external demands for quality control. As a user of the facilities, the director (M) is willing to take his chances with a new computer alternative. What he also strives for is an effective use of the facility through automatisation. He has a problem that needs to be solved. Considering himself as an internal customer he is able to override the clash of roles because of his position. As a non-specialist, he is nevertheless deeply involved in technical aspects of innovation.

Concerned actors

Let us recapitulate. The group included two staff managers responsible for buildings and operational security (A) and quality (Q), one member of the purchasing department (SI) and, finally, the controller assigned to the support department (STE). All these actors have been concerned in their capacity as a functional expert. However, they have performed minor roles in the development process. The quality manager (Q) is responsible for quality audits according to the rules of ISO standards. He has initiated the project when informing about pending environmental legislation and quality standards about maximum emission levels for particles and certain chemical substances in exhaust fumes. Thus, he has voiced the need for the implementation of functioning control systems to warrant that engines conform to the *new environmental and quality standards*. This signals reflexivity on a strategic company level.

The manager responsible for buildings and operational security (A) has been obliged to make estimations of investment needs, so that buildings comply with new functional demands (e.g., water and air temperature maximum variation levels). He complains. 'Our technicians have an ambition level that is much too high as to technical specifications . . . they are not realistic in their assumptions. This has made it difficult for me . . . I have to argue my great investment needs with top management.' A has problems of translating *new functional demands* into financial figures of investment to ascertain *correct economic reports*. The latter element is introduced by the controller (STE). He gives a current status report. 'The new budget signals that we keep the total project costs within the limits of appropriations for the years concerned.' This fact, as a follow-up, further connects with the struggle of the manager of the entire support department (ST) to arrive at *a good decision*.

Finally, one member of the purchasing department (SI) tells his story of how he had to switch supplier at the last minute to be able to hang on to the process.

> I entered the picture relatively late . . . the supplier had already been
> selected. My task was to negotiate a reasonable price and to finalise
> a good contract. There was panic . . . the supplier had already reserved
> equipment for us several months ago. I gave the supplier an order
> over the telephone. The next day I was told that another supplier had
> been selected . . . because a PC-based system had been decided on. I
> felt very bad. I told him that we had chosen a cheaper alternative.

SI summarises. 'This is not really the way we want to operate in the
purchasing department. We have to be informed at a much earlier date.
It must be important to use our purchasing expertise.'

This material signals the use of *purchasing expertise* as an additional quality
element to make the supplier extra responsive to buyer needs. However,
the importance of complementarity between roles during innovation is
obvious. Staff experts come into the process at certain phases and rely on
the input from others. At critical moments, they may be obliged to change
their normal duties and procedures. This is clearly shown by the embarrassing situation for the actor SI. Flexibility has to be used.

All in all, the case has illustrated how an internal customer (user) together
with a PC system supplier helped a test manager to push through a decision to scrap the old system and install a new certifying facility. The struggle really concerned expertise and power in the support department.
However, the operators of the facility were deeply involved during installation to make it user-friendly. Let us finalise this chapter by summarising
our theoretical conclusions in terms of our suggested model of the innovation process.

Discussion

What can we learn from quality models of innovation processes?

First, let us recapitulate. We have modelled internal innovation processes
from the vantage point of quality dimensions that participating actors introduce. In contrast to traditional modelling methods no direct relations
between activities and actors are mapped. Instead, the underlying quality
dimensions (elements) are elicited from the unrestricted verbal materials
of actors. By means of interpretation, carried out as a dialogue with the
respondents, these verbal materials are reduced to key expressions, which
are intended to capture the quality dimension. These expressions, then,
are the empirical counterparts to what we have termed quality elements.

These latter can subsequently be organised into a quality model by cate-
gorising actors as to their *centrality*. It is important to note that the model
as such is conceptual in nature. It should therefore be interpreted from
dialogues with involved actors, rather than from prescriptive assessments
by design or planning unit employees. Consequently, its usefulness depends
on the skills of the researcher in engaging in open-ended dialogues with

respondents by letting them become participants in the research process as joint interpreters. Let us now summarise our findings about strategic reflexivity and its four sub-components: interaction, roles' complementarity (between roles) and innovation management in terms of our model.

The whole point with this quality model is to illustrate how the innovation process is organised from what is crucial, namely novelty. Degree of novelty has been termed quality elements. By capturing how sequential steps of novelty (quality) emerge during innovation work we suggest that the most important part of innovation *per se* is evidenced. Innovation does not arrive in ready forms. Rather, it emerges from micro-level interactions between actors who collaborate, albeit with different expectations and roles. The quality model also depicts how actor roles become intertwined in terms of how they are able to insert quality elements in to the process. Each role will have a distinct infusion of quality which is the outcome of the function of the role in the process.

By differentiating between quality elements, we can also analyse *the degree of complementarity between actors and their roles*. In the case there was an overlap developing between the roles of the actors in the two computer camps. 'Struggling actors' had a different view of what kind of quality was necessary. At the heart of the issue was the struggle for the keeping of the role of the computer expert. User-friendliness of new systems could, and did, replace that role.

'Instrumental actors' had complementary roles, that of the user (or internal customer) and supplier. They were able to support innovation by focusing on the practicalities of operating the system. Other less central roles, such as those of the 'concerned actors' depicted in the quality model, were also rather complementary. Each of these latter roles added important input and output elements to the innovation process. *New environmental and quality standards*, for example, initiated the whole project. The other elements were infused as practical and administrative consequences of innovation.

Theoretically, we could argue that less overlap between roles should lead to less role conflict. Notwithstanding, that conflict can also be creative in the sense that it may mobilise employees into more directed activity. Too much conflict, on the other hand, can cause inertia. New roles were created in the course of the project. Complementarity between roles should facilitate collaboration if properly managed. However, as illustrated by the embarrassing position of the purchasing officer who had to cancel the order, expertise has to be co-ordinated in order to be effectively used.

The non-hierarchical nature of the innovation process is clearly depicted by the model. It categorises actors as to their centrality in taking part in innovation work; or rather as to how complete an account of the entire process they have been able to muster. Centrality understood in these terms is different from management position or rank in the organisation. Individual employees or other actors can have a very central and important role without having a defined decision authority. For instance, the manager of the entire support department (ST), to which all struggling actors reported, had less to say about the course of events than the operating manager of the test facilities STP. The former was caught in the middle of the struggle and strove for a good decision and work climate. On the other hand, it is striking how focal actors in the case (such as STP) have had to cope with problems of getting enough resources for their 'underground' activities. The director of the profit centre (M) expressed it as follows: 'The PC alternative grew up like a pirate venture in the firm.' By forming alliances between departments and external experts he drove the innovation process his way.

His account is a good example of how an actor *can renegotiate strategies of innovation by changing his role and the expectations of other roles.* He breaks away from the rule-governed system in a systematic way. This is clearly an exemplar of strategic reflexivity in action. There seems to be at least two managerial roles involved. One is to drive the new alternative forward (STP). Another is to arrive at a position of consensus by open-ended discussions. In this way, the manager of the support department (ST) tries to control 'chaos'. Finally, he has no alternative but to go along with the new PC-alternative.

Type of interaction between the actors involved in the innovation process *is also analysable* by the quality model. By depicting how quality elements hang together in time (not displayed here) we are able to study the necessary outcomes of these interactions. As such the model becomes a blueprint for what has been achieved. The 'nodes of novelty' are exposed as a time-dependent structure of quality outcomes.

The gist of the innovation process seems to be derived from a few essential, quality elements. The aim of attaining minimal quality levels along respective dimensions must succeed. For example, bundles of reoccurring activities have taken place during installation work of the PC-system. Practically speaking, each type of control signal is tested and is logged in a stepwise fashion until maximum tolerance levels have been complied with.

Another, and final, interesting feature in the case is the dependence of external competence. The representative's (L) responsiveness and expertise in adapting a process-based control system to a different industrial setting made him a winner. Against the will of the experts, novices in the computer field introduced a new kind of system competence. Knowledge is infused from the outside, rather than secretly developed inside an R&D department. Assimilating new external knowledge becomes a crucial part of innovation. We conclude by arguing that strategic reflexivity can be mapped and become clearly visible by building quality models of the innovation process. As such, it becomes a blueprint for what has been achieved in terms of novelty, the very essence of innovation.

Part IV

Policy

Strategic reflexivity can be said to evolve around certain collectively organised contexts that makes it possible. Such contexts can be sectorial, for example part of the service sector, or the public sector, but it can also be embodied in policy processes. In order to understand the character of such contexts at the systemic or societal level, other approaches compatible with strategic reflexivity become relevant. In this section, two such complementary approaches are presented and discussed in relation to strategic reflexivity, namely a systems approach emphasising the role of policy in the 'innovation system', and a social constructivist approach, emphasising a broader sectorial or network oriented conceptualisation of innovation. As examples of these approaches, this part will discuss the role and character of the EU innovation policy and the construction of digital cities in local public administration.

Susana Borrás develops the idea that innovation, understood as strategic reflexivity, is a negotiatiated process, which is partly a result of collectively organised contexts where such negotiations are undertaken. Innovations are socially and institutionally embedded within a given context, and this context is partly shaped by public actions. Public action has, according to Borrás, a double role in the innovation process by setting the context where innovation takes place and by inducing a *collective* reflexive process where a self-perception of the innovation system/context can emerge. Public action in relation to innovation can, in Borrás' framework, be boiled down to four distinct functions that set the context and stimulate self-perception of the innovation system, namely regulation, enhancing the generation of knowledge, creating inter-connectedness, and reducing socio-territorial disparities. Borrás goes on to analyse the 'innovation policy' of the EU as an example of such public action.

Birgit Jæger examines innovation at the broader sectorial level in public administration by means of a case-study, the development over the past years of new public, electronic services to the citizens called 'digital cities'. Jæger applies a social constructivist approach to understand how the digital cities have come into being. The social constructivist approach implies that technology is understood as shaped or constructed through social processes. Jæger uses the concepts of configuration and representation to further qualify the approach. Configuration is a way to describe how the user is built in to (configured in) the technology. Where 'configuration' primarily deals with the actual design of the technology, 'representation' deals with the conceptions that the designers use as the basis for the configuration of the technology. Jæger then goes on to analyse representation and configuration problems in the construction of digital cities. She also discusses 'social learning' processes involved in this process. Finally, she attempts to link social constructivism to the concept of strategic reflexivity arguing that the two approaches are compatible in some respects, though social constructivism is a broader approach, while strategic reflexivity tends to take its starting point in the firm or the organisation.

13 Recent trends of EU innovation policy

A new context for strategic reflexivity

Susana Borrás

Introduction

The theoretical consideration which developed in the first chapter of this book argues that innovation is a strategic and reflexive process. This has an important effect on the way in which public policy is defined in analytical terms. Strategic reflexive theory maintains that innovators operate in a largely unstructured, uncertain and rapidly changing environment. They operate strategically in order to position themselves in relation to competitors and customers, and they operate reflexively by anticipating the results of their own actions in the market. This means that an innovation has been negotiated internally within the firm and that it has involved external partners, and the consumers in the market place. However, innovations should not be understood as a mere aggregation of these individually negotiated decisions. They are *partly the result of a collectively organised context where such negotiations are undertaken.* In other words, innovations are socially and institutionally embedded within a given context, and this context is partly shaped by a set of public actions.

The main argument of this chapter is that public action plays a double role in the innovation process. First, by setting the context where innovation takes place through important public functions like regulation, the generation of knowledge, the enhancement of connectivity, and the pursuit of balanced socio-economic development in the territory. And second, by inducing a *collective* reflexive process where a self-perception of the innovation system/context can emerge. The empirical study of this postulate will be an examination of the way in which the recent trends of the EU's innovation policy are actually re-shaping and Europeanising the context for innovators in the old continent.

This chapter proceeds as follows. First, it will consider, generally and theoretically, the functions of public action in the innovation process. Second, the chapter will briefly examine the recent trends in EU involvement in line with the analytical framework of the previous chapters. In this sense, it will respectively address these two questions: what kind of instruments does the EU currently have in this field; and how have they recently evolved in line with the new innovation policy paradigm? And, to what extent does this set of initiatives constitute a reference-point producing reflexivity of the innovators at European level? The conclusion will sum up the arguments and discuss whether and how the new EU-initiatives are really creating a new context for innovation in Europe.

On public action's role: shaping the context for innovation

The introduction to this book develops the notion that the innovation process is essentially the fruit of the strategic reflexive actions of innovators. Strategic because innovators position themselves *vis-à-vis* competitors' alternative innovative solutions/products. And reflexive because this positioning takes place through each innovator's self-understanding in relation to the challenges to be faced. This means that innovation is essentially a social process characterised by the strategic and reflexive nature of actors' interactions. However, the innovation process cannot be understood as a mere aggregation of these interactions, mainly because they take place within a rich but wider political, economic and social context. This is the question about the embeddedness of social action in institutions which is so central to institutional economics. In this sense, institutional economists have successfully developed the notion 'systems of innovation', capturing the embeddedness of the innovation process in an institutionalised context. Whereas initial research focused on 'national systems of innovation', subsequent studies have examined 'regional', 'local' and even 'European' systems.[1] In all of them, there are formal as well as informal institutions which delineate the particularities of such systems.

However, the governance of the system is an issue that has been relatively neglected by the economic studies. In particular the role of public action in the innovation system has generally been seen as subordinate, and mostly in normative terms, when formulating the 'policy recommendations'. Fortunately, this perception is currently being reversed by some theoretical work which analyses the different models of innovation systems (Amable *et al.* 1997), the governance of regional systems of innovation (Cooke *et al.* 2000), and which has a theoretical focus on the role of government in system analysis (Steen 2000). In what follows, I will develop a schematic approach to this matter.

System governance is a wide concept, whereas public action is a more restricted one. By system governance I refer to the constitutive and interactive elements that define (and redefine) the nature and performance of the system. By public action I refer to the set of political goals and instruments that public actors develop in order to enhance the innovative/ technological competencies of the administered territory. In this sense governance extends well beyond the mere scope of public action, including, more generally, elements of collective and social action through all sorts of formal *and informal* institutions as steering mechanisms. Public action is narrower in the sense that it focuses on the role of public actors, more specifically in the sphere of public policy.

Public action has four distinct and important functions in the innovation system, namely, regulation, enhancing the generation of knowledge, creating inter-connectedness, and reducing socio-territorial disparities. *Regulation* is the state's use of rules to govern the economy and society. As such, it has a paramount role in the innovation process as it defines the nature and the legal framework for the interactions between innovators. Regulatory areas that are concerned with innovation are: competition regulation (guaranteeing and controlling fair competition), company law (firms' statutes and operations), financial regulations (stock market, banking sector), taxation regulations, industrial property rights (especially intellectual property rights such as patents), safety, health and environmental regulations, and bioethics (such as clinical practices and rules about human cloning). All these regulatory fields affect the innovative activity in many different ways by distributing responsibilities, granting rights and obligations, defining roles, and setting standards and limits. In any case, the most important question for this regulatory setting to be a supportive base for innovation is that it produces legal certainty. That is, a regulatory system that sets clearly and enforces effectively (and fairly), social interactions. And this is especially important for the innovation process, given its uncertainty and its fragmented and conflict-based form of social interactions.

Enhancing the generation of knowledge is another key function of public action, through the public support of science and of the educational system. Traditionally, science has had a privileged status in the eyes of policymakers. However, today there is the general acknowledgement that innovation is related more broadly to all sorts of knowledge production and learning processes. This means that public action focuses on the overall generation of knowledge, and not just on the generation of *scientific* knowledge. In addition to the idea that we live in a 'knowledge-based economy', building the knowledge competencies of the system in its widest sense is a key element in the strategy for growth and competitiveness (Lundvall 1999). Emphasis is therefore placed on the following issues, namely, elaborating a new 'social contract of basic science' that acknowledges basic science's key role in the development of human resources and competencies (Martin and Salter 1996), the ability of educational and training systems to produce life-long learning individuals, and the wider enhancement of the knowledge base and the learning abilities of organisations (Dalum *et al.* 1992; Storper 1996).

Creating connectedness in the system, is the third function of public action especially since the 1980s. This is so because the relative lack of synergy and interaction between the actors in the territory has been perceived as a bottleneck to innovativeness. Thus, the political goal has been that of fostering the connections between different types of innovative organisations such as firms, scientific researchers, consumers, laboratories and so forth. The policy instruments have mainly been networking programmes, bridging institutions, technology transfer offices and several forms of databases.

The reduction of socio-territorial disparities is a key function of the welfare state and has an important relation to the innovation process, especially for the less developed territories. The efforts to reduce the gap in less developed regions have traditionally been carried out through public investment schemes. In recent years technology and innovation matters have started to become an important aspect of such development schemes. Specific initiatives have been the creation of testing/certification laboratories, of bridging institutions, of start-ups incubators, special fiscal advantages for industry, the improvement of research conditions for universities or even the relocation of public research facilities. Not that they can all claim to be success stories, not even to have definitively reversed the bad economic situation for those territories. But public action in all these areas has played an active catalysing effect for the (scarce) innovation that is found there.

As can be clearly seen, the four functions of public action are not mutually exclusive. That is, the function of generating knowledge might well be related to some instruments which are used to enhance the connectedness of the system, or even the reduction of disparities. Yet, allowing a certain degree of overlap, these functions do seem to encompass the role of public intervention in shaping the context for innovation.

Moreover, taken together, these functions of public action might have an important and overall effect, namely, one of *inducing a collective reflexive process from the innovators*. All forms of public action involve communication. The diverse public functions and instruments I have mentioned above in relation to innovation do have this communicative dimension, even if they were not explicitly designed for that purpose. And in this communicative dimension there is the key for awakening collective reflexive process or system self-perception, which is a sort of a meta-function of public action. This meta-function is important in two senses. First, because this collective reflexive process creates individual expectations for innovators about how the context is organised, and about what can he/she obtain from it. These initial expectations, which might be based on trust and confidence, influence individual courses of action. Take, for example, the regulation of intellectual property rights. The collective expectations about the costs/benefits and about the strong/weak patent system in a given state/region, define patenting propensity and how individual firms manage their intellectual property assets. Second, it is important because this collective reflexive process induces innovators to act politically, that is, to engage in political discussions about how to reorganise the functions of public action. Following the same example, generalised dissatisfaction with a lax intellectual property rights' system mobilises firms and industrial organisations pressuring for a stricter regulation system, with more protective effects and higher penalty rates for infringements.

To sum up, public action supports the strategic reflexivity of innovators in two ways. First, by setting the context where innovation takes place through important public functions like regulation, the generation of knowledge, the enhancement of connectivity, and the pursuit of a balanced socio-economic development in the territory. These four functions structure the general framework where the innovation related negotiations happen, and hence the consistency, efficiency and transparency of this framework are of utmost importance for the innovation process. Second, public action supports innovation in a more indirect way by inducing a collective reflexive process where a self-perception of the innovation system emerges. Individual innovators act accordingly to their collectively induced expectations about the system, not only in their innovation activities but also when they act politically trying to re-shape the forms of public action.

EU innovation policy in the 1990s: re-thinking public functions

Along with most national policies, the EU policy of the 1990s has moved from the 'technology policy' rationale of the 1980s to a 'innovation policy' in the mid-1990s (Sanz and Borrás 2000). The 'green paper on innovation' signalled a turning point for the EU involvement in this direction (Commission 1995; Commission 1996), because it re-packaged a new agenda into some existing EU initiatives, and because it provided a course for further EU public action in the 2000s. The new innovation policy is based on the rationale that public action has to deal with the overall context where innovators operate, rather than just with strategic RTD programmes.

How are EU public actions articulated today? How are they setting a new context for innovation in the 2000s?

Regulation and innovation in the EU. The EU has a significant regulatory framework which is the fruit of political decisions and negotiations institutionalised gradually in the so called 'aquis communautaire'. EU regulations are an important element of the context where innovators operate, mainly because these regulations formalise the multiple forms of interactions between economic actors. In other words, they set the 'rules of the game' for economic interaction (also innovative activities). More generally though, regulation and regulatory reform have been related to growth and competitiveness in the EU (Galli and Pelkmans 2000). The regulatory areas most directly connected with innovative activities are mainly competition policy, intellectual property rights and bioethics; and in an indirect way, environmental law, health and safety regulations, public procurement and company law.

Starting with *competition policy*, the EU has long regulated and controlled fair competition in the single market. It has done so by punishing abuses of dominant position and by prohibiting market distorting agreements between firms. Technological co-operation and intellectual property rights-related agreements between firms are allowed, but the Commission and the European Court of Justice (ECJ) ensure that there are no abuses (Korah 1996). The regulation of *intellectual property rights* (IPRs) is another crucial regulatory area for innovating firms. IPRs have traditionally been regulated at national level, but the growing internationalisation of the markets has stimulated the desire for the protection of these rights at trans- and inter-national levels. The European Court of Justice has for long controlled the abuse of IPRs as barriers for intra-EU trade, in an extensive jurisprudence concerning IPRs' relationship with EU competition policy (Anderman 1998). Besides, the EU has started to regulate IPRs from the mid-1990s. It has created important EU IPR legal figures such as trademarks and copyrights, and at the time of writing is very close to creating Community utility models, designs and the Community patent. These legal figures allow the appropriation of knowledge and innovative production, and hence are key elements for the creation of a true context for innovation (OECD 1997).[2] *Bioethics* is the third crucial regulatory instrument of the EU, as it sets limits to the innovative enterprise. These limits are based on values that are socially and politically negotiated by considering what is and what is not a desirable effect of science and technology. The EU has recently regulated some of these aspects, such as the cloning of humans, clinical and research practices, and the controversial normative about the patentability of biological inventions.[3] Table 13.1 summarises the relationship of these three EU regulatory areas with the innovation process.

Besides competition, IPRs and bio-ethics, the EU regulations about environmental law, health and safety regulations, public procurement and company law also affect the innovative context though in a less direct way. This can be by setting some compulsory environmental and health protection measures that force innovators to search for new technical solutions (Clinch 2000); by creating a European market for public tenders that forces competition among the firms' technological competencies, as in procurement policy (Edquist and Hommen 1998); or by creating a simplified legal framework for firms' legal structures that can enhance more inter-firm co-operation, as in company law.

Enhancing the generation of knowledge in the EU. Public action directed towards the generation of knowledge is generally focused on scientific research and education. Concerning the first, the EU has had the so-called RTD framework programme since the mid-1980s, through which it finances scientific projects of trans-national nature provided that they have a 'European added value'.[4] EU allocations function as a complement, and not as a substitute, of national public financing of research. In the late 1990s, the allocations of the framework programme did not grow as spectacularly as they did in the late 1980s and early 1990s, but they still represent the third largest part of the EU budget.[5] Most recently, the Commissioner, Busquin, has launched the initiative of a 'research area', which aims at

Table 13.1 The relationship of three important EU regulatory areas with the innovation process

Regulatory area	Relationship with innovation process
Competition policy	Prohibits and controls unfair competition: • Technological co-operation between firms is allowed, but the Commission-ECJ controls that there is no market abuse • Technology transfer/Intellectual property rights-related agreements should not hinder internal market regulations
Intellectual property rights	The EU grants property rights in the form of: • Trade marks • Copyrights • Designs (under decision stage) • Utility models (under decision stage) All of them allowing the appropriation and exclusive exploitation *at EU15 level* of the knowledge produced by the innovator
Bioethics	Limits scientific and innovative practices on the basis of socially and politically negotiated values. The EU has recently started to regulate in this area, e.g. regulating clinical practices in humans, or prohibiting human cloning

strengthening the transnational dimension of scientific research carried out in Europe (Commission 2000). An important novelty of the 1990s in relation to all public funding of research has been the close examination of the additionality question of public support, or to the extent to which public support is not used to substitute expenditures which would be made anyway by private innovators. In other words, a careful analysis is done of the boundaries between financial incentives and plain subsidies. This generalised attitude at national level has also been maintained for the EU Framework Program, as reflected in the V Framework Programme (1999–2002) (Luukkonen 2000).

Concerning education and training systems, the EU has undertaken major initiatives in this respect. Along with the premises of the 'knowledge-based economy' and 'learning economy', and with the concerns about job generating economic policies, the issues of human capital, education, training, mobility and life-long learning have acquired a front-line position. For this reason, the EU programmes in areas of training and mobility established in the 1980s (mainly Commet-Erasmus), were reinforced in the 1990s by a broader perspective on these issues in the pursuit of competitiveness and innovation. Successive Commission initiatives in the field, in the form of communications and white/green papers give accounts of this political broadening of the issue.[6]

Creating inter-connectedness is another function of the EU involvement in innovation matters that has developed significantly in the 1990s. Even if the initiatives are difficult to grasp, mainly because they are not organised systematically, they do show the clear willingness of the Commission to enhance the internal connectivity of innovators in the EU context. Besides the networking effects of the RTD and education programmes, the Commission has also fostered networking through more indirect channels. These include the task-forces panels organised by Edith Cresson concerning 'the car of the future' or 'Aeronautics', the arrangement of thematic conferences where scientists and industrialists meet to discuss specific topics like 'SMEs and innovation in Europe'. Another strategy to enhance inter-connectivity has been building new institutions, like the Office for the Harmonisation of the Internal Market (OHIM) in Alicante, or like the intellectual property rights help desk with information-communication about these rights in Europe. Furthermore, networking has been a continuous interest of EU-action, with initiatives like the network of technology transfer offices, or the network of innovation relay stations (which provide advice and channel contacts between innovators in Europe). Recently, emphasis has been placed in diffusion-dissemination of research results (through an impressive database system) and in encouraging venture and high-risk capital in Europe.[7] More discrete than other EU initiatives, this myriad of EU incentives for informal and formal inter-connectedness at European level might in the long run have a strong impact.

Reducing socio-territorial disparities in the EU context has since the 1980s been centred on regional economies and articulated through the regional policy of the Union (also called 'cohesion policy'). This policy has operated through different development programmes, which support a large range of public investments expected to boost economic growth (Bachtler 1998). In the 1990s, these premises were partly reoriented, introducing more innovation-related initiatives and budgetary posts in the large programmes of the structural funds (Landabaso 1997; Borrás and Johansen 2001). This implies a relative rapprochement of EU regional and RTD policies in a convergent objective: reducing the technology gap *within* the Union (Sharp 1998). Moreover, the future enlargement eastwards has also helped profile politically the issue of disparities and diversity.

EU innovation policy in the 1990s: inducing collective action

The alliance between the Commission and some large industrial firms in the early 1980s succeeded in persuading national governments to create the RTD framework programme (Sharp 1991; Peterson 1992; Cram 1997). The most powerful communicative-political device in the persuasion process was the notion of the 'European technology gap' *vis-à-vis* the US and Japan. A gap which Europe had problems in narrowing in the absence of collective (EU) initiative. Some economists have recently made the criticism that the 'gap' has been politically exaggerated in some respects but not in others (Pavitt 1998) (Tijssen and van Wijk 1999). Nevertheless, this notion continues today to be the motto for EU involvement in RTD programmes *and* innovation-related initiatives (in the new version of 'innovation gap' rather than 'technology gap'). In any case, the reason for the continuation of this vision is mainly that the 'Europeanness' it appeals to seems to be quite uncontested. In other words, there seems to be a generalised acceptance among political elites that there is a need for European collective action in technology/innovation issues.

Indeed, the reflexive effects of the new EU innovation policy strategies need to be more systematically analysed, especially in the late 1990s–early 2000s. A contrasting picture seems to be emerging. On the one hand, the borders of the EU RTD framework programme are becoming increasingly blurred by an extensive participation of non-EU firms/researchers and by an increasing institutional co-operation between the EU and other non-EU pan-European research organisations like ESF, CERN or ESA[8] (Borrás 2000). In this sense, the definition of 'us' and 'Europeanness' in strict EU terms is becoming increasingly unclear given the globalisation of knowledge production (Strange 1998; Väyrynen 1998). On the other hand, there are signs of an emerging EU self-referential point in other innovation-related areas. This is most clearly indicated by the active interest of European industrial organisations and researchers in the area of EU intellectual property rights and bio-ethics.

Conclusions

The European Union has undoubtedly been engaged in creating a new context for European innovators in the 1990s. It has done so by re-shaping or launching new initiatives along four different functions of public action, namely, regulation, generating knowledge, creating inter-connectedness and reducing disparities. The limits of this chapter did not allow us to get closer to the pitfalls and hurdles that each of these public actions initiatives still face, to discuss success/failure stories about EU involvement in innovation, nor to examine the innovators' perception about the 'European system'. However, in general, the EU of the 1990s–2000s is clearly committed to this matter, and has an extensive strategy in this sense – something which was not the case in the 1980s.

It is difficult not to finish by pointing to further research questions that result from the focus on 'strategic reflexivity' in the EU context. Particularly interesting in this regard is the question about the 'systemness' effect of the re-shaped and increased EU involvement in this matter. If the 'systemness' of a system is defined by a collective reflexive self-perception, by a certain (high) level of interaction between system actors and by a set of differentiated public action functions, the European Union might be close to being defined as a European System of innovation of a post-national nature. Yet, this tag should not hide that the EU innovation policy is indeed a very ambitious political project due to the large diversity in terms of socio-economic structure, of growth patterns, of economic institutions, of innovation traditions and of the science–state relationships among its member states.

Notes

1   Caracostas and Soete 1997; Braczyk *et al.* 1998.
2   See Table 13.1.
3   Directive 98/44/EC, OJ L 213, 30/07/1998, p. 0013–0021.
4   The most relevant scientific fields are: information and communication technologies, life sciences, energy research, industrial technology and new materials.
5   The V Framework Programme (1998–2002) has been allocated a total 14.9 bECUs, increasing 'only' 12.5 per cent to the previous one (1994–8). This limited increase (when compared to the geometrical increments of previous framework programmes) has to do with the generalised budgetary constraints imposed by the Maastricht criteria of the EMU project.
6   The most relevant documents-action plans are: the white paper on 'Teaching and learning: towards the learning society' COM(95) 590; 'Learning in the information society'; action plan for a European education initiative (1996–8), COM(96) 471; the Commission communication 'Towards a Europe of knowledge', COM(97) 563, and the green paper: 'Living and working in the information society: people First', COM(96) 389.
7   The diverse initiatives about this issue can be found in 'financing innovation' web-page:
    http://www.cordis.lu/finance/home.html
8   ESF: European Science Foundation, CERN: European laboratory for particle physics; ESA: European Space Agency.

14 Innovations in public administration

Between political reforms and user needs[1]

Birgit Jæger

Introduction

Innovations do not only take place in private enterprises, they also take place within public administration which has otherwise had a tradition for introducing few, and then only incremental, changes. Planning in public administration used to consist, more or less, of automatically extrapolating the budget. For many different reasons this tradition has changed dramatically within the last few decades. In the 1970s and 1980s it became clear that large scale societal changes (due to such things as an increasingly overstitched economy, growing unemployment, the introduction of information technology, and an increase in international relations) demanded new solutions. In other words, there was a need and a political demand for innovation in and development of the public administration.

There is a huge body of literature which analyses the extensive development of public administration during this period (e.g. March and Olsen, 1989, Bogason, 2001, Rhodes, 1997) which I will only briefly refer to in this chapter. Generally speaking, one can say that there has been an introduction of market mechanisms into public administration. Several reforms, often referred to as New Public Management reforms, have had as their objectives to improve efficiency, effectiveness and accountability in the public administration. Amongst other things, the reforms have included initiatives such as: the decentralisation of local administration; introduction of user boards; improved systems of budgetary control; a greater reliance on financial incentives; privatisation of former public services and increased competition between public and private services (Boston, 1999). Altogether the reforms have resulted in a breakdown of the traditional hierarchy of public organisations. Instead, new networks have appeared where the borders of the organisations are not clear at all, and where public and private organisations are operating side by side (Andersen *et al.*, 1999)

A great number of these reforms have been possible because of information and communication technology (ICT). An example of this is the decentralisation of economic competence to local public agencies in the mid-1980s which were to a large extent made possible because local public agencies became equipped with, at that time, new personal computers. ICT has also been the tool for rationalising a lot of administrative routines and procedures such as budget control, the keeping of accounts and so forth: just as it has been the backbone in many organisational changes in public administrations. There is also a huge body of literature describing the role of technology in the change of public administration (Andersen, Friis *et al.*, 1999).

Starting with this brief sketch of reforms in public administration, this chapter will focus on one example of innovation in public administration where reform and technology walk hand in hand. The innovation consists of the development of a new public electronic service to the citizens called Digital Cities. This chapter will focus on the development of digital cities in Denmark as well as in the rest of Europe. It is based on a European study, called the SLIM study,[2] in which seven European digital cities were analysed.[3] Digital cities will be defined just as the historical and political background will be lined up.

This chapter draws heavily on the work undertaken by SLIM, thus it is based upon the same social constructivist perspective as that employed in the SLIM study. In other words, technology is considered to be a social construction. That is to say that technology is created by the actors that are involved in the development process, and that it does not embody any 'given values' or characteristics. In other words, the technology is formed in a network of different actors which have different interpretations of the technology.[4] The Danish and other European digital cities will be analysed in the light of this theoretical perspective. The analysis is carried out on the basis of a few selected theoretical concepts. I have also chosen to discuss the digital cities based on the theoretical concepts of representation and configuration. Next I will analyse the social learning processes in the digital cities. I conclude on the basis of the analysis and consider the findings in the light of the strategic reflexivity.

Digital cities

Definition of the digital city

The term 'digital city' emerged in the early part of 1994 when one was established in Amsterdam (De Digitale Stad – DDS), (Lieshout, 2000). As with all new terms, there is not yet a 'definitive definition' of what a digital city actually is. One interpretation sees the digital city as a concrete city that, often via a development project, is equipped with an electronic infrastructure. In other words, according to this interpretation it is the infrastructure itself that serves as the basis for the definition.

Other interpretations place greater emphasis on the content of the electronic infrastructure. In this case, a distinction can be made between two interpretations: those digital cities that are 'grounded' and those that are 'non-grounded'. The 'grounded' interpretation sees the digital city as a concrete, existing city that offers citizens' access to electronic information and communication. The information in such a digital city will be narrowly tied to the existing city. An example of such a digital city is the Copenhagen Base. The 'non-grounded' interpretation sees the digital city as an electronic city where the 'citizens' that share the same interests can meet and exchange ideas and experiences. According to this interpretation, there is no necessary connection to a concrete city. An example of such a digital city is DDS in Amsterdam, where only 22 per cent of the users in 1996 were actually resident in Amsterdam (Lieshout, 2000).

The term digital city has yet to gain widespread usage in Denmark. Mostly there is talk about local authorities' websites or city-net. Recently, it seems as though there is a distinction emerging between these terms. The usage of the term 'websites' is restricted to public homepages on the web. While the term 'city-net' is used to describe websites announced under the name of the town including both public authorities, private enterprises and all kinds of civil society associations and other organisations.

In the SLIM study we used the term digital city to describe all these kinds of electronic services. Our point of departure was that we wanted to study innovations in public administration. Thus we did not use the term digital city in relation to services where the public authorities did not participate even though the web site might be using the name of a city.

The background for the digital city

Even though the term digital city has not been definitively defined it is nonetheless possible to trace the vision behind the virtual local community. Visions of 'The Wired Cities' (Dutton *et al.*, 1987) emerged in the 1960s in the USA, which on many points are reminiscent of contemporary notions concerning the digital city. While the technological foundation for these ideas was the spreading of cable television, the vision concerning the possibility of local communities possibly employing local television to facilitate debate and disseminate information is not significantly removed from contemporary visions concerning the virtual local community. These visions can also be rediscovered in the ideas concerning local television and local radio in many European countries throughout the 1960s and 1970s.

It was already suggested in the 1970s that information technology could serve as the basis for the realisation of the visions described above. While the Internet's cornerstone was already laid in 1972, and was called ARPANET at the time, many years passed before it was made available to the public. For that reason, the first attempts at establishing an electronic local net were primarily based on bulletin boards (BBS).

Emerging policies for entering the information society

The vision of the virtual, local community is part of a broader political vision of an information society. Thus it is possible to find the vision in the American NII-initiative (the National Information Infrastructure), which Clinton and Gore introduced shortly after they assumed the presidency in 1992. In the American case there was not only talk of local electronic networks, but also of global networks for the benefit of both commercial and private interests. A European working group under the leadership of the German EU commissioner Martin Bangemann followed up the American NII initiative. This work resulted in the so-called Bangemann report, which was published in 1994 (Bangemann *et al.*, 1994).

The Danish government took up the Bangemann report's challenge and established a governmental committee to clarify the manner in which the information society was to develop. This resulted in the so-called Dybkjær report, which was published in 1994 (Danish Ministry of Research, 1994). The Dybkjær report concerns itself with many different aspects of the information society. As regards the electronic services provided by the public authorities, the report concluded that the entire public administration – state, county and municipality – ought to be connected within a comprehensive service net. This net should provide better services to citizens and commercial enterprises, rationalisation windfalls, as well as a more open decision-making process. The principle was to result in the following initiatives:

- All public institutions were to establish an electronic mailbox.
- All ministries were to develop a plan for electronic communication.
- Public institutions were to develop electronic self-service systems and bulletin boards.
- 'Statens Information' (the state information service) was to establish an electronic map for public instances and institutions as well as an electronic bulletin board with press releases and publications from the institutions of the state.

At that time it was not clear which electronic infrastructure should be used. The Dybkjær report advocates that the electronic infrastructure for this service net should be Diatel.[5] But already two years later in 1996 it was made clear that the Internet was to serve as the electronic infrastructure.

> The connections between the public nets is to be based on Internet technology, which has become the de facto standard for communication in the open nets and is increasingly used as the basis for communication in the internal nets of the commercial enterprises, so-called 'intranets'. As such, the goal is to create 'Internet functionality' between the public nets in a public intranet.
>
> (Danish Ministry of Research, 1996, p. 55)

The Dybkjær report was followed by a series of action plans.[6] In 1999 a new governmental committee was set up to revise the Danish ICT policy. This work resulted in the report entitled 'The Digital Denmark' (Det Digitale Danmark) where five fields were pointed out to be the leading areas in the further development. One of them was the digital public administration. Thus there is even more focus on digital innovations in public administration today then ever before. To create a chronological overview I have listed some important events preceding the digital cities in Table 14.1.

Table 14.1 Important events preceding the digital cities

1960s	The 'wired cities' idea emerges in USA – visions about using local TV and local radio for communicating within and between local communities
1970s	The first 'wired cities' projects are established in USA and Japan
1972	First demonstration of ARPANET (the Internet's predecessor)
1974	Establishment of the first 'Community network' in Berkeley, USA (community memory)
1980	Establishment of the first local community's BBS in USA (Old Colorado City)
1986	Establishment of the first 'Free-net' in Cleveland, USA
1989	Establishment of PEN (Public Electronic Network) in Santa Monica, USA
1990	First implementation of World Wide Web Establishment of the first cities' home pages in USA (e.g. Palo Alto)
1993	Publication of 'NII: Agenda For Action' in USA
1994	Publication of the Bangemann report in Europe Publication of the Dybkjær report in Denmark
1995	The Danish government's first plan of action was announced The first websites of local authorities were established in Denmark
1999	Publication of the Digital Denmark report in Denmark
2000	217 out of 275 Danish municipalities have their own website. Every 14 counties have their own website. It is not registered how many governmental agencies have their own website but the impression is that it is a great deal

European digital cities in a theoretical perspective

For many years the study of technological innovation has focused on the research and development process – often in laboratories and big companies. The development of technology was understood as a linear process which took place in different stages such as: invention, innovation, and diffusion. Just as technology was understood as a determining society (Smith and Marx, 1994). This way of understanding technology and the development of technology has been questioned by another theoretical approach, where technology is understood as being shaped or constructed by social processes. From studies within the social constructivist framework we have learned that it is not appropriate to describe the development of technology as a linear process. Studies have shown that there are still processes of innovation going on even in the stage of diffusion (Bijker, 1995) and concepts like innoffusion have been developed to describe this overlap of innovation and diffusion. This new understanding has led to a need for studying technology in a broader context where social processes both in the lab and in the everyday use of technology are drawn into the analysis.

The SLIM study was based on this social constructivist understanding of technological development and innovation. Thus the users' interaction with the electronic service is regarded as a part of the innovation process. It is probably always important but when the product is an electronic, public service it seems especially important to draw in the users' way of appropriating the technology. In this section the innovation of the European digital cities will be discussed against the background of some of the theoretical approach developed in the SLIM study. The topics that I have chosen to discuss here build on the theoretical concepts of *configuration* and *representation*.

The concept configuration was originally developed by Steve Woolgar in an article where he described the manner in which the conception of the coming users of a PC were 'built in' to the technology itself (Woolgar, 1991). Implicit in the concept is the sense that the conception of the coming users has direct influence on the physical design of the technology. Conversely, the physical design of the technology can serve to determine which users will ultimately use it, as well as the way they do so.

The SLIM study has employed the concept with respect to the configuration of the users. This has led to the concept 'representation'. In other words, the two concepts are closely linked to one another. Where 'configuration' primarily deals with the actual design of the technology, 'representation' deals with the ideas that the designers (and here we are talking about all of the actors that are involved in the development of the technology) use as the basis for the configuration of the technology.

The SLIM study investigates the representation of the users, the city and the technology. Representation of the users indicates the image that the designers have of the potential users and their use of the technology. Representation of the city indicates the image that the designers have of the city. Representation of the technology indicates the image that the designers have of the technology.

Representations of users

A common characteristic of the European digital cities is that the users/citizens (there is no distinction made) have not had any particular influence on the design of the digital cities. The Copenhagen Base (CB) is one of the examples where the users have been involved to the greatest degree. In the

CB the internal users from the city's administration were involved as a part of a user group from the very beginning. However, this involvement of users is only partial, as the members in the user group only represented the employees at the Copenhagen City Hall. In other words, it is merely a small fraction of the 46,000 city employees that are represented. All of the employees in the city's institutions, schools, libraries, centres for the elderly, etc. are not represented in the user group, even though they also are considered to be among the internal users of the CB.

The external users – in principle all citizens in the City of Copenhagen – have been involved via an evaluation that was conducted in 1997. The evaluation particularly emphasised the investigation of the users' evaluation of the CB. These external users could also have been included in other ways. For example, one could, for a period, have tested the CB as a social experiment, but the web-master estimated that it would be difficult to achieve the necessary political support for it. Today the external users primarily express themselves via email, which they send directly to the web-master. He indicates that the users' response is of considerable value, for which reason one must be surprised over the fact that the evaluation's recommendations to involve the external users more directly in the continued innovation have not been implemented.

When the users themselves are not directly involved in the process of innovation, the designers must design the technology on the basis of their own hypotheses concerning the users. This then is what I refer to as the representation of the users. This representation of the users can to some extent be observed in the concrete design of the technology. Concerning the case study of the CB Bastelaer *et al.* state: 'To conclude, the external user of the Copenhagen Base is a local inhabitant, well-trained and moderately equipped, mainly interested in city administrative information and a rather passive consumer of this information more than as an active citizen' (Bastelaer *et al.*, 2000, part 3, p. 28).

This representation of the users is not general for all seven of the digital cities. For example, in Amsterdam the users are considered not so much as consumers of information but more as citizens who themselves produce information and create social and electronic links to other users. This is reflected in, among other things, the fact that DDS provides opportunities for the individual user to lay out various things (information, matters of debate, etc.) in the common room that other users can express their reaction to.

When considering the seven European cases as a whole, it is clear that it is the technical expertise and competence that have been made the highest priority, whereas the users' wishes and competence do not appear to have played any significant role. The representation of the users that the designers employ as their basis when they are developing the concrete technology is far from explicit in all cases. Often the designer himself is not even conscious of it. It is also rare that the representation is tested, for example via the involvement of user groups or social experiments. If new information concerning the users turns up in the middle of the course of development, the representation of the users can suddenly shift. The different groups of actors that are involved in the development of the technology can also have different representations of the users. All of this contributes to a situation in which one can rarely trace a coherent, conscious representation of the users that serves to guide the development over a longer period.

Representation of the city

As mentioned above, the SLIM study also worked with the representation of the city. It is a significant difference whether the city is exclusively employed as a metaphor in the user interface, as in the case of DDS in Amsterdam, without being intended as a service for a concrete, physical city – in other words, an example of a non-grounded digital city. Or whether it is a grounded digital city that is based on a concrete city and its residents as in the case of the CB and the CCIS in Edinburgh. The concept of representation of the city therefore raises the question: where are the city limits of a digital city?

The representation of the city can to a certain extent also be observed in the concrete design of the digital city. CB and Périclès are examples of representations of cities that are dominated by public authorities and grant very little space to the common citizens. By contrast, DDS is an example of a representation of a city where the citizens play a much more active and direct role.

Representation of technology

Just as in the case of the representation of the users and the city, technology is also designed on the background of a representation of the technology itself. The general tendency in the SLIM study's seven cases is that the designers of the digital cities have quite a conservative perspective on the technology. They regard it primarily as a means of distributing information, while the multimedia many-faceted potential – in terms of content, presentation, interactivity and user interface, etc. – are touched upon to a very limited extent.

It has gradually become commonly accepted that there is a special 'Internet culture' that is often described on the basis of hacker circles and as a kind of ICT-grassroots movement. Within this culture, the Internet is perceived as a kind of free space where everyone will have the opportunity to speak. In this free space experiments are often conducted concerning the potential and limits of the technology, and many new applications have been discovered in this way. It was actually in this manner that it was discovered that information technology could be utilised for communication.

When the first French videotex system was introduced (in the mid-1980s), it was exclusively designed for the dissemination of information. It was a group of French hackers who discovered that they could communicate with one another by using the technology in a manner that had not been foreseen by the designers (Feenberg, 1992). This discovery was promptly developed into the technology and became a significant factor contributing to the great success that videotex enjoyed in France (Jæger, 1995).

Another group of actors experimenting with the technology's potential is artists. For example, a number of artists have drawn on video technology to create experimental video-installations. Such artistic experimentation with different media has also led to the discovery of many new applications.

The idea behind the first version of DDS in Amsterdam was fostered in a co-operative venture between precisely these two groups of actors, hackers and artists (Lieshout, 2000), and the first version was distinctly influenced by the so-called Internet culture. As DDS in Amsterdam was the first digital city in the world, it is no exaggeration to say that the very idea of a digital city can be accredited to these groups that dared to experiment with the technology's limits and potentials.

In the meantime, the Internet culture did not continue to dominate the development of the digital cities. When the first experiment with DDS had been completed it was decided that it should continue on commercial terms and artistic experimentation was halted. Despite the fact that DDS in Amsterdam had been the first digital city, the Internet culture was not carried over to the other European digital cities. The rest of the cases (including the CB) have, as already mentioned, been influenced by a relatively conservative representation of the technology.

Configuration of technology and users

While representation deals with the conception of the users and technology (and the city), configuration deals more with the physical design – the configuration itself – of the technology and the users. One can say that the designers can force their representation of the technology and the users upon the users on the basis of the manner in which the technology has been configured. The representation of the users has a bearing on the configuration of the technology, just as the configuration of the technology has a bearing on the configuration of the users.

The designers often find themselves in a dilemma when they are to configure the technology. They must decide whether it is to be configured in a manner so that all types of users can utilise it or whether it is to be configured to certain user groups, well knowing that many of the potential users will be excluded. The result of the former strategy is often that the technology does not reach any users at all. In the second version of DDS they tried to configure the technology for as many users as possible. DDS was required to survive on commercial terms, so the designers wished to be able to reach the broadest possible number of potential users. This led to a development in which emphasis was placed on technical finesses, but where the users were excluded from the design. This also led to a configuration where many of the possibilities for communication – which had been included in the first version – were omitted (Lieshout, 2000).

It is remarkable that the Danish telephone companies (this was prior to the establishment of TeleDanmark) had precisely the same experience in the mid-1980s when they wished to develop a videotex system on commercial terms. In order to be able to attract as many potential users as possible they attempted to develop a system that was compatible with all of the existing forms of hardware and software. This led to such a technically complicated system that it was only those users with a very high degree of computer competence that were able to use it (Jæger, 1995).

The seven case studies conducted of the SLIM network have revealed that there are different means of configuring the users:

- The user interface has great influence on the configuration of the users. The manner with which the information is presented and the metaphors that are used all configure certain user groups and to a certain use of the technology.
- The language used is also of great significance for the configuration of the users. The decision only to use Danish in the CB serves to exclude all potential users who do not speak Danish, just as the formal language attracts certain users and repels others.
- The content and types of information also have significance for the configuration. For example, DDS offers free email. This attracts some other users than CB's offering of city information.
- The rules in the digital city also play a role in configuring the users. In the CB there are rules that do not permit groups with politically extreme views to speak. In other words, these users are excluded beforehand. No rules can also have the exclusion of potential users as a consequence. Such an example occurred when TeleDanmark introduced the so-called 900-telephone numbers at the beginning of the 1990s, where one could establish telephone-based services that the users were to pay. TeleDanmark did not wish to make rules for the use of the 900-numbers. They perceived it to be an open marketplace where all could develop services. From the outset there was a boom in the supply of services offering telephone sex. Almost immediately, most Danes came to equate the 900-numbers with telephone sex and an increasing number of workplaces and private subscribers set up barriers for the use of these numbers. 'Nice citizens' did not wish to be associated with telephone sex. In the end, TeleDanmark had to give up their hope that a wealth of different telephone-based services would emerge.
- The opportunities for access naturally also have significance for the configuration of the users. In DDS, for example, computers are not

made accessible to the public. In other words, users who do not have a computer in their home or at work are excluded from the digital city in Amsterdam.

- Introductory courses or other forms of training in the use of the digital city can also play a role in configuring the users.
- The users are also configured via advertising. As a result of the significant power that advertising has in our society, there is reason to believe that the manner with which the service is presented in advertisements plays a role in determining who ends up using the service. If all advertisements are directed at a certain target group, other target groups will presumably not be attracted as users to the same degree.
- Instructions and technical guides also play a role in configuring the users. Complicated technical guides speak to a different user group than user-friendly instructions directly presented on the screen with small, funny figures.
- Finally, experiments, trials and test periods can also play a role in determining the users. By conducting these, the designers can achieve a more precise and detailed representation of the users, which possibly better corresponds to the users experience of her own needs and desires.

Social learning and innovation

As the title of the SLIM study (Social Learning in Multimedia) indicates, the concept of social learning has been of fundamental significance to the project. The reason for our interest in social learning is rooted in the theoretical understanding of innovation. If innovation and the development of technology is taking place in the interaction between different actors some kind of learning or reflexivity must be at work. Innovation is not only a question of creativity but also a question of reflexivity where the actors learn from earlier or other actor's experience. Our reason for referring to social learning processes is to point out that we are not focusing on individual learning processes. The SLIM study has exclusively been interested in investigating societal, collective, learning processes. One way to analyse social learning processes was to study social experiments with ICT. Based on descriptions from the eight participating European countries, the SLIM study wanted to analysis social learning, which resulted from experiments. In this section I will discuss social experiment as a pathway to social learning. After this I will discuss different theoretical understandings of social learning.

Experiments as learning processes

Throughout the rest of the world, Denmark is commonly regarded as a country where there has been a great deal of social experimentation with

ICT, and where ordinary citizens have had the opportunity to try out the technology in their own everyday situations. At the end of the 1980s, for example, an experimental programme was conducted by the state in which sixteen experiments using information technology received support. These were placed in different peripheral regions (Cronberg, 1990; Cronberg *et al.*, 1991; Jæger *et al.*, 1990). Both TeleDanmark and Kommunedata have conducted their own experiments and in conjunction with the government's strategy for enabling Denmark to become an information society, ten so-called 'spear-head' municipalities were selected to test different possible designs for a local information society.[7]

In all of these experiments the actors learned a great deal about technology, about the manner in which one organises the local experiments, as well as precisely how their local community can best use the technology. One of the most significant conclusions of the evaluation of the sixteen experiments in the 1980s was that some important learning processes had taken place in the local communities (Cronberg, 1990). The experiments had not succeeded in reversing the negative development in the peripheral areas they had declared as an objective. In that sense they had failed but they had succeeded with regard to social learning.

These experiments with information technology have attracted significant international interest. There is, therefore, reason to believe that those involved in the development of the Copenhagen Base have learned from these specific Danish experiences. However, in response to the question as to whether they had learned something from the earlier Danish experiences, both the CB's web-master and a representative from the user group replied negatively. Neither of them had ever even heard of the Danish experiments with ICT! And that despite the fact that many of them have been evaluated, such that there are research-based evaluation reports available to the public. In these reports it is additionally possible to find descriptions of precisely the same problems that the CB itself has struggled to solve.[8]

This discovery is somewhat remarkable and serves to cast light on problems related to the EC's official policy concerning the dissemination of experiences. The EC expects that research projects such as the SLIM study will result in a mapping of 'best practice'. In other words, a mapping of where one has progressed furthest in such areas as the use of multimedia in the public administration and a description of the conditions that were present in the successful incidents. The idea is then, that the best practice can thereafter be transferred to other cities, countries, or local communities.

The SLIM study serves to sow doubt as to the wisdom of this strategy in the EC (Williams, *et al.*, 2000). First, it is rare that one can identify any clearly successful criteria. Second, comparative studies indicate that conditions are rarely sufficiently identical such that it is possible to merely transfer such experiences and expect to achieve the same successful result.

Third, one can learn just as much from a failure as from a success. Therefore there are good reasons to disseminate the experiences from the projects that failed to reach the mark. With a theoretical basis such as the SLIM study's, where local relations, particular cultural conditions, the local actors' different abilities, etc. are attributed decisive significance for the concrete development, it is also difficult to argue that one can merely transfer experiences from a successful project to a different local context.

Even though one can not merely transfer aspects of a success story to different contexts, the experiences from one project can be drawn upon when considering other projects and put to use where it is considered relevant on the grounds of the social context. The SLIM study also demonstrates that there are several European digital cities that have been inspired by DDS in Amsterdam (particularly in the case of DMA in Antwerp). Just as we know that there has been an exchange of experiences between the members of a European organisation named Telecities.

Another means of sharing such experiences can be achieved by making staff changes. A former project leader will bring her experiences with her to a new project, whereas a project leader without such experience must begin from scratch. The CB is an example of a project where they started from scratch, without any contact with similar projects for which reason they have had to accumulate their own experiences. Conversely, there have been many learning processes in the period that has passed within the auspices of the CB itself, and today there is a willingness to share these experiences with others. The web-master told of how other cities that are interested in creating a municipal website/digital city have visited Copenhagen and heard about how they have done it.

Theoretical understandings of social learning processes

Processes of social learning are mostly processes where one can later ascertain that learning has occurred, even though it is not possible to place a finger on when and how it occurred. Because it is empirically difficult to trace the learning process it is also difficult to understand theoretically.[9] If one wishes to investigate such learning processes, it would seem obvious to resort to organisation theory.

On the basis of classic studies of organisations' behaviour, organisation theoreticians such as Levitt and March (1988) define organisational learning as the construction of experience-based routines that thereafter determine the organisation's behaviour. When they speak of routines, they are referring to rules, procedures, strategies and technologies, but also the way of thinking, paradigms, cultures and knowledge. According to such an understanding, learning can be observed via the modification of the organisation's routines. In other words, the organisation must begin to act differently than it has in the past in order to be able to say that a process of learning has occurred.

Other, more cognitively oriented, theories concerning organisational learning such as Björkegren (1989) define learning as a modification of the predominant means of thinking in the organisation. However, such a modification of patterns of thought will also manifest itself in the organisation's actions, in which case the learning process can again be observed via a modification of practices.[10]

These (and other) theories about organisational learning can be useful when we investigate the learning processes that have occurred, for example, in the City of Copenhagen in conjunction with the CB. It would seem obvious that learning has occurred on the basis of the fact that the city has modified the means by which it relates to the citizens. Some of the organisation's internal routines have been modified, as indicated by the fact that two persons have been employed to co-ordinate CB, just as some of the employees in the different departments have been assigned the task of updating the information in the CB. But these modifications are of a relatively superficial nature in terms of their significance for the organisation. There has been no significant change in terms of the patterns of thinking concerning the manner in which the city should services its citizens. Even though the citizens can electronically communicate to the city, they still receive their response via a standard letter on paper. And even though the web-master predicts that the CB will come to change the city's perspective towards the citizens, this change is not yet evident in any form of a modification of routines.

Within organisation theory there is a conception of two different types of learning. Argyris and Schön (1978) write about single-loop and double-loop learning, and Björkegren (1989) writes about learning through assimilation and accommodation. Learning through assimilation (or single-loop learning) is said to occur when the organisation adjusts its perception of the world without radically altering its image of the world. Existing knowledge in the organisation contributes to this adjustment. Assimilation maximises the inner stability because it confirms that the organisation's fundamental image of the world is in order. At the same time, however, assimilation involves a danger that if an organisation is continually able to confirm its image of the world, it then becomes difficult to recognise when the demands of the world around it actually call for a change in this perception of the world around them.

Learning via accommodation (or double-loop learning) occurs when the organisation fundamentally changes its perception of the world. Such processes create entirely new knowledge. This type of learning is traumatic and threatening because it exposes holes and contradictions in existing knowledge, which necessitates the reconstruction of the organisation's knowledge before assimilation can again take over. This type of learning does not occur particularly often, as it involves significant instability in the organisation, but when it does occur it maximises the learning due to the fact that it leads to entirely new perspectives when an organisation modifies its fundamental perception of the world.

These two concepts can be employed to describe the City of Copenhagen's learning in conjunction with the CB as learning via assimilation. The organisation has come to perceive some new signals from the world around it: the citizens have a greater need for information concerning the city and the technology that has been developed provides new possibilities for the dissemination of such information. Just as the political signals have pointed to the necessity for innovate electronic, public services. By developing the CB, the city has changed some of its routines such that they fit the adjustments to the image of the world better. But there is no talk of a fundamental modification of the city's sense of reality.

Organisation theory can contribute to an understanding of the social learning processes that occur in connection with the development of a digital city. But there is still something missing. As mentioned above, the SLIM study builds on a theoretical perspective that regards the development of technology as a process, which proceeds within a network of different actors. Some of these actors are organizations (e.g. the City of Copenhagen and IBM), while other actors are single individuals (e.g. the users) or loosely organised groups and associations. Organisation theory focuses, by definition, only on organisations. While the world around it might well play a role in the organisation's development the theories about organisational learning are inadequate in terms of describing the interplay between organisations and different actors in the world around them. We lack therefore a theory that can serve to explain the learning processes in a network perspective.

In a quest after such a theory I have become acquainted with the so-called 'Theory of Situated Learning' (Lave and Wenger, 1991). This theory is based on a study of apprenticeship in different cultural contexts. The empirical studies indicate that learning occurs as a social interaction in a community of practice. This leads to a perspective on learning as a dynamic that occurs within a specific historical and cultural context.

Lave and Wenger describe learning as a process where an apprentice is granted access to a specific community of practice. Such a community of practice consists of experienced individuals – old-timers – such as journeymen and a master craftsman, as well as other apprentices. The community's knowledge is not the exclusive domain of the master craftsman; rather, it is embedded in the relations between the persons and the activity in the community. On that background, access to the community is decisive for the learning process. On the basis of this access, the newcomer is granted a legitimate position in the periphery of the community of practice. Lave and Wenger describe how the apprentice thereafter moves towards the centre of the community of practice via legitimate peripheral participation, and after some time joins the ranks of the old-timers. Through this process the apprentice acquires knowledge and skills, just as her identity is shaped such that she becomes a member of the community. Lave and Wenger describe this process in the following manner:

From a broadly peripheral perspective, apprentices gradually assemble a general idea of what constitutes the practice of the community. This uneven sketch of the enterprise (available if there is legitimate access) might include who is involved; what they do; what everyday life is like; how masters talk, walk, work, and generally conduct their lives; how people who are not part of the community of practice interact with it; what other learners are doing; and what learners need to learn to become full practitioners. It includes an increasing understanding of how, when, and about what old-timers collaborate, collude, and collide, and what they enjoy, dislike, respect, and admire. In particular, it offers examples (which are grounds and motivation for learning activity), including masters, finished products, and more advanced apprentices in the process of becoming full practitioners.

(Lave and Wenger, 1991, p. 95)

This approach to learning shifts the focus from the individual to the collective, social level. It is no longer a question of 'pouring information into the apprentice' but rather, a question of access to a community of practice. Nor is it any longer a question of cognitive processes in the mind of the single individual, but rather, a matter of participation. This shift of perspective makes the concept of community of practice very important, which Lave and Wenger define as:

A community of practice is a set of relations among persons, activity, and world, over time and in relation with other tangential and overlapping communities of practice. A community of practice is an intrinsic condition for the existence of knowledge, not least because it provides the interpretive support necessary for making sense of its heritage. Thus, participation in the cultural practice in which any knowledge exists is an epistemological principle of learning. The social structure of this practice, its power relations, and its conditions for legitimacy define possibilities for learning (i.e., for legitimate peripheral participation).

(Lave and Wenger, 1991, p. 98)

Lave and Wenger do not consider the community of practice to be a static entity. The community changes over time. These changes are a result of the fact that the newcomers have a different interpretation of the community of practice and its future. Conflicts can arise between the old-timers' and newcomers' respective interpretations of the community, which in turn can contribute to developments in the community.

If we wish to understand the story of the CB on the basis of situated learning theory, questions are immediately raised concerning where we actually find a community of practice. When the political decision was made that the City of Copenhagen was to develop the CB, there was no community of practice in the city that was capable of immediately assuming the task. The first challenge, then, was to develop such a community of practice. This was achieved via the establishment of the users group that was created to develop the CB. It was within this community that a common interpretation of what an electronic service such as the CB should actually include was developed, as well as a common representation of the users and the technology. The participants in the first users group were recruited from each of the city's six magistrate departments. Most of them were appointed on the grounds that they were the ones from the department that had demonstrated the greatest interest in the technical challenges. This meant that the common interpretation that was created in the users group was quite focused on the technical possibilities in relation to creating the CB.

When the CB's web-master was hired to develop the base he was not at all an expert in electronic services. He was merely a librarian with an interest in the subject but without any practical experience. When he contacted IBM in Denmark and they engaged in a joint venture to lay the first version of the CB on the Internet, he became a newcomer in IBM's technical community of practice. He learned a lot from this community of practice: about technical standards and solutions to technical problems. But he did not learn anything about the structuring of the city's information searching possibilities or proper language in a public electronic service. These elements were not a part of the community of practice that he had been granted access to in IBM.

The web-master's new technical expertise was much in harmony with the users group's interpretation of the CB. But after some time the group was supplemented with information officers from the different departments. These newcomers to the community of practice brought new approaches to the CB with them. They insisted that the matter of the structure of information and the language in which it was presented in the CB should be taken up for discussion in the group. This led to conflicts within the group concerning the interpretation of the CB, the representation of the users and the representation of the technology. These conflicts have not yet been concluded, but their result can have decisive significance for the configuration of the CB.

An explanation for the fact that neither the web-master nor the other participants in the community of practice surrounding the CB have drawn on the experiences from previous Danish experiments with ICT can be found on the basis of the theory of situated learning. None of them have had access to a community of practice where these experiences have been a part of the collective knowledge. Such communities of practice function today within TeleDanmark and Kommunedata (both of whom have been directly involved in the experiments), in the local communities where the experiments occurred, and within the group of researchers that have evaluated the various experiments. The City of Copenhagen has not had contact with any of these communities of practice when they were to begin developing the CB. It has obviously not been of any help that there were diverse evaluation reports standing on the shelves of the public libraries when there was nobody to make those involved in the CB's development aware of their existence.

While somewhat superficial, this analysis nevertheless serves to demonstrate that the theory of situated learning can uncover other elements of the social learning process than the theories of organisational learning provide occasion to. However, this analysis does not provide the background which is necessary to be able to conclude whether situated learning theory is also the answer to the need for a theory of learning processes: a theory that is capable of capturing the interaction between various types of actors who engage in a network surrounding the development of a digital city. There is a need for a more precise definition of what lies within the concept of 'community of practice'. Can this concept be used in relation to large, complex networks such as those involved in conjunction with the development of a digital city, or can it only be used in relation to more limited communities such as those that exist within an organisation?

While Lave and Wenger base their approach on the apprenticeship system that occurs in a workshop, their examples serve to demonstrate that the theory of situated learning can also be employed in other contexts.[11] But it is still an open question as to the limits for the theory's validity. In which concrete situations does it make sense (or not) to employ the theory of situated learning?

Conclusions and discussion

In this chapter we have analysed an innovation in public administration called the digital city. The analysis based on the theoretical concepts of the representation of users, technology and the city and the concept of the configuration of the technology suggested some interesting perspectives on how innovation is occurring when a new public service is designed. Discussing the public authorities' use of ICT in the dialogue with the citizens is not just a matter of describing the most successful examples and encouraging the public authorities in other cities to return home and do the same. We must first understand the processes that lay behind both the successful projects as well as the projects that ultimately are judged to be fiascos. First, when we have a greater understanding for the background it will be possible to utilise this knowledge towards the further development of public authorities' use of ICT. The theoretical reflections will therefore be regarded as attempts at proceeding towards a greater understanding of the background processes. The concepts of representation and configuration thus contribute to a greater understanding of the fact that the development of electronic services is not merely a matter of technology. There are both conscious and unconscious processes that are in play in the concrete design of the service.

The theories concerning organisational learning brought us a step forward on the path towards achieving understanding of the processes around the development of the digital city. But as we saw, these theories lack a perspective according to which the learning process is understood as something that occurs within a network between various actors, for which reason I have drawn on the theory of situated learning. This theory can serve to uncover other elements in the learning process, though it is as yet too early to conclude to what extent it is the answer to our need for a theory about learning in a network perspective. Further theoretical consideration is necessary, particularly concerning the concept of the community of practice and the manner with which this concept can be related to the network perspective. Just as a clarification of the situations in which the theory can be employed is necessary.

In conclusion I will point out that the comparative perspective has proven itself to be very fertile in this study. By comparing the development of digital cities around Europe, various aspects of the projects have come to light that would not have been possible to analyse against the background of a study of an individual case. By comparing them to one another it becomes apparent that those things which were taken for granted in certain situations, such as in conjunction with the development of the CB, are not necessarily relevant in all contexts. The comparative perspective makes it apparent that the Danish tradition for public administration and the Danish perception of democracy has had a decisive influence on the concrete design of a digital city in Copenhagen. These conclusions have helped make us a little wiser concerning the processes that are in play during the concrete development of technological tools in the dialogue between public authorities and the citizens.

Similarities and differences between the social constructivist and the strategic reflexivity approaches

If we compare the social constructivist approach with the strategic reflexivity approach, outlined in the introduction to this book, we can find both similarities and differences in the perspectives. The similarities are based on the same wish to go beyond the understanding of innovation as a linear process. Both perspectives point out that innovation is a much more complicated process just as both perspectives stress that we are talking about processes of social interaction among human actors.

There is also a similarity in the concern of reflexivity and social learning. In the SLIM network we have been discussing the two concepts and asked ourselves whether the concepts were describing the same phenomenon just from different points of view and different scientific traditions. We did not arrive at an unambiguous answer but it is clear that the two concepts are closely related.

These were the similarities in the two approaches but I also see some significant differences. The strategic reflexivity approach is considered from the point of view of a private firm. This point of departure leads to an understanding of innovation where the objective is to capitalise upon the outcome of the innovation. Actually, this is a part of the definition of innovation in the Schumpeter tradition. The social constructivist approach operates with a much more broad definition of innovation. In this perspective, innovation does not have to result in commercial outputs. Actually, a streamline of thought within this approach is working with user innovation when they appropriate the technology in their homes (Silverstone and Hirsch, 1992). In these situations the innovation does not lead to a commercial product but to a better utilisation of the artefact by the users.

Using the firm as a starting point also results in a great deal of emphasis on the manager's role in managing the firm. The strategy referred to in strategic reflexivity is the strategy of the manager. I could be tempted to say that this is just a kind of strategic management where the new strategy is reflexivity. If reflexivity or learning is a strategy for owning more money, the management of the firm need to have a strategy for how to be reflexive!

In the social constructivist approach managers' strategies for maximising the income is just one feature out of many, which is seen as having influence on the innovation processes. Of course, this strategy can play a crucial role in some situations but in other situations it only plays a small role or maybe no role at all. The social constructivist approach does not leave firms, mangers and economic strategies out of the analysis. Actually, there is a strong tradition within this perspective of working in this field.[12] But the difference is that the social constructivist approach does not a priori give these features precedence over other features in the analysis. It is up to the empirical analysis to reveal whether this is of importance in the case in question or not.

The point of departure in the firm naturally implies a focus on economic theory. The social constructivist approach is also inspired by different economic theory. But beside economic theory this approach is also inspired by more traditional sociological and cultural theory. This gives the social constructivist approach a much broader scope than the strategic reflexivity approach.

A final difference between the two approaches is the understanding of organisations. The strategic reflexivity approach operates with the firm as a well-defined organisation. The social constructivist approach also operates with organisations but not only as firms. This chapter is an example of a social constructivist analysis of a public organisation, beside the fact that the social constructivist approach sees organisations as a part of a network where the border between the organisation and its surrounding is not clear and stable. What is also characteristic of the social constructivist approach is that the network is established between different kinds of actors. Some of the actors are organisations (typically firms or public agencies) while others are single individuals (typically users) or organisations from the civil society (e.g. environmental movements, hacker societies and so forth). By letting these other types of actors into the analysis the social constructivist approach opens up the possibility for a much broader analysis than the strategic reflexivity approach does.

To conclude: the two approaches are based on ontologies which are in many ways similar. This, altogether, makes it possible to have a discussion across the two perspectives. But the differences are also significant. The point of departure in the firm of the strategic reflexivity approach leads it in other directions than the social constructivist approach. The strategic reflexivity approach is engaged in the development of a theory for understanding innovations in private firms with the objective to earn more money. The social constructivist approach is engaged in the development of a theory for understanding innovation in a broader social science perspective. This implies that the strategic reflexivity approach has a much more focused, limited, scope than the social constructivist approach.

Anyway, despite the differences, the shared wish to go beyond the linear understanding of innovation and develop an understanding of the complex and multifaceted process of innovation could be a reason to let the two approaches inspire each other. I am quite sure that both approaches could be fertilised by some of the insights of the other approach.

Notes

1 This chapter is based on a case study described in Bastelaer *et al.* (2000). Parts of this chapter have earlier been published in Danish.
2 Social Learning in Multimedia (SLIM) was a research project running from 1996 to 1999. It was funded by the European Commission DG XII Targeted Socio-Economic Research (TSER) programme. The project was led by the University of Edinburgh and had participants from Belgium, Denmark, Germany, Ireland, The Netherlands, Norway and Switzerland.
3 The SLIM study described the following digital cities: The Digital City ('De Digitale Stad' – DDS) in Amsterdam (the Netherlands); Craigmillar Community Information Service (CCIS) in Craigmillar (Scotland); Digital Metropolis Antwerp (DMA) in Antwerp (Belgium); Freehouse 2000 ('Frihus 2000') in Frederikstad (Norway); Geneva-MAN in Geneva (Switzerland); Périclès in Namur (Belgium) and the Copenhagen Base (CB) in Copenhagen (Denmark).
4 A more comprehensive description of the social constructivist perspective in technology studies can be found in Bijker, 1995 and Jæger, 1995. That of the SLIM-project's theoretical perspective can be found in Williams, final report, as well as in Van Lieshout, *et al.* (2001).
5 Diatel was a Danish on-line service similar to American services such as American Online and Compuserve. Diatel consisted of a co-operative venture between TeleDanmark, two national newspapers (Politiken and Jyllandsposten), a bank (GiroBank) and Kommunedata, a large company specialising in supplying software to the Danish municipalities. Diatel was introduced in March 1995 as the electronic network of the future in Denmark, but it was closed in February 1996 on the grounds that it could not compete with the Internet. The story of Diatel is told in Jæger, 1995.

6 A more detailed description of the Danish national ICT policy is to be found in Jæger and Hansen, 1999.

7 A comprehensive presentation of experiments with ICT in Denmark can be found in Jæger and Hansen (1999). An analysis of the experiences with social experiments in the countries that the SLIM-project covers can be found in Jæger *et al.* (2000). The lessons from the social experiments are discussed in Jæger (forthcoming).

8 In Kommunedata's trials with videotex in 1989–91 they also encountered problems related to the structuring of the information. They, like others, had structured the information on the basis of the administration's structure, which resulted in a situation in which the users could not find the right information. The problem and the attempts at solving it are described in Jæger and Rieper, 1991.

9 At least if the methodology is inductive, which is often the case with a social construtivistic approach. I will not go further into this discussion here.

10 These theories are presented and discussed in Jæger (1994).

11 One of their concrete examples is drawn from the organisation Alcoholics Anonymous. They write that AA involves a learning process where alcoholics learn to be non-drinking alcoholics.

12 For an overview of the field, see Williams and Edge (1996).

Bibliography

Acs, Z. and Audretsch, D. (1990a) 'Small firms in the 1990s', in Acs Z. and Audretsch, D. (eds) *The Economics of Small Firms*, Norwell MA: Kluwer.

Acs, Z. and Audretsch, D. (1990b) *Innovation and Small Firms*, Cambridge MA: MIT Press.

Alavi, M. and Joachimsthaler, E.A. (1992) 'Revisiting DSS implementation research: a meta-analysis of the literature and suggestions for researchers', *MIS Quarterly*, 16 (1): 95–116.

Albert, S. and Bradley, K. (1997) *Managing Knowledge: Experts, Agencies and Organizations*, Cambridge: Cambridge University Press.

Ældrekommisionen (1980) *Aldersforandringer – de ældres vilkår*. Ældrekommissionens 1. delrapport, København: Ældrekommissionen.

Amable, B., Barré, R. *et al.* (1997) 'Diversity, coherence and transformations of innovation systems', in R. Barré, M. Gibbons and S. J. Maddox (eds) *Science in Tomorrow's Europe*, Paris: Economica.

Anderman, S.D. (1998) *EC Competition Law and Intellectual Property Rights: The Regulation of Innovation*, Oxford: Clarendon Press.

Andersen, E.S. (1994) *Elements of Evolutionary Economics*, London: Pinter.

Andersen, H.S., Jensen, L., Jæger, B., Sehested, K. and Sørensen, E. (1999) *Roles in Transition! Politicians and Administrators between Hierarchy and Network*. Research Paper no. 7/99, Roskilde: Roskilde University.

Andersen, K.V., Friis, C., Hoff, J. and Nicolajsen, H.W. (1999) *Informationsteknologi, organisation og forandring – den offentlige sektor under forvandling*, Copenhagen: Jurist- og Økonomforbundets Forlag.

Ansoff, H.I. (1965; 1987) *Corporate Strategy*, London: Penguin.

Ansoff, H.I. (1982) *Strategic Management*, London: Macmillan.

Antonelli, C. (1995) *The Economics of Localized Technological Change and Industrial Dynamics*, Boston: Kluwer.

Antonelli, C. (1996) 'Localized technological change: new information technology and the knowledge-based economy: the European evidence', Oslo: SI4S project, European Commission (DG XII), TSER programme.

Antonelli, C. (1999) 'The evolution of the industrial organisation of the production of knowledge', *Cambridge Journal of Economics*, 23 (3): 243–60.

Argyris, C. and Schön, D. (1978) *Organizational Learning: A Theory of Action Perspective*, Reading, MA: Addison-Wesley.

Arieti, S. (1976) *Creativity: The Magic Synthesis*, New York: Basic Books.

Arrow, K.J. (1969) 'Classificatory notes on the production and transmission of technical knowledge', *American Economic Review*, 59 (2): 29–35.

Arthur, W.B. (1989) 'Competing technologies, increasing returns, and lock-ins by historical events', *Economic Journal*, 99 (394): 116–31.

Atkinson, A.B. and Stigliz, J.E. (1969) 'A new view of technological change', *Economic Journal*, 79: 125–53.

Augsdorfer, P. (1996) *Forbidden Fruit*, Aldershot: Avebury.

Aydalot, P. (1986) 'Trajectoires technologiques et milieux innovateurs', in Aydalot, P. (ed.) *Milieux innovateurs en Europe*, Paris: GREMI.

Bachtler, J. (1998) 'Reforming the structural funds: challenges for EU regional policy', *European Planning Studies*, 6 (6): 645–64.

Badham, R. (1993) 'Introduction: new technology and the implementation process', *The International Journal of Human Factors in Manufacturing*, 3 (1): 3–13.

Bagella M. and L. Becchetti (eds) (2000) *The Competitive Advantage of Industrial Districts*, Heidelberg: Physica Verlag.

Bangemann, M. *et al.* (1994) *Europe and the Global Information Society: Recommendations to the European Council*, Brussels: EC.

Barki, H. and Hartwick, J. (1994) 'Measuring user participation, user involvement, and user attitude', *MIS Quarterly*, 18 (1): 59–82.

Barras, R. (1986) 'Towards a theory of innovation in services', *Research Policy*, 15: 161–73.

Barrett, P. (1996) 'The good and bad die young', *Marketing*, (1): 16–17.

Basberg, B.L. (1987) Patents and the measurement of technological change: a survey of the literature', *Research Policy*, 16 (2–4): 131–41.

Bastelaer, B. van, Henin, L and Lobet-Maris, C. (2000). *Villes virtuelles. Entre Communauté et Cité. Analyse de cas*, Paris, L'Harmattan.

Bastick, T. (1982) *Intuition: How We Think and Act*, Chichester: Wiley.

Bateson, J.E.G. (1992) *Managing Services Marketing*, second edition, London: The Dryden Press.

Baudrillard, J. (1983) *Simulations*, New York: Semiotext.

Becattini G. (1989) 'Riflessioni sul distretto industriale marshalliano come concetto socio-economico', *Stato e Mercato*, April: 111–28.

Beck, U. (1986) *Risikogesellschaft*, Frankfurt: Suhrkamp.

Bemelmans, T.M.A. (1991) *Bestuurlijke informatiesystemen en automatisering*, Deventer: Kluwer Bedrijfswetenschappen/Stenfert Kroese B.V.

Bessant, J. and Caffyn, S. (2001) 'An evolutionary model of continuous improvement behaviour', *Technovation*, 21 (3): 67–77.

Bessant, J. and Francis, D. (1999) 'Developing strategic continuous improvement capability', *International Journal of Operations and Production Management*, 19 (11): 1106–39.

Bessant, J. and Rush, H. (1995) 'Building bridges for innovation: the role of consultants in technology transfer', *Research Policy*, 24 (1): 97–114.

Bessant, J. and Tsekouras, G. (2001) 'Developing learning networks', *A.I. and Society*, 15 (2): 82–98.

Best, S. and Kellner, D. (1991) *Postmodern Theory-Critical Interrogations*, London: Macmillan.

Betænkning (1947) *Betænkning afgivet af det af Arbejds- og Socialministeriet den 24. maj 1946 nedsatte udvalg angaaende husmoderafløsere*, København: Arbejds- og socialministeriet.

Bijker, W. E. (1995) *Of Bicycles, Bakelites, and Bulbs: Toward a Theory of Sociotechnical Change*, Cambridge MA: MIT Press.

Bilderbeek, R. and Den Hertog, P. (1997) *The New Knowledge Infrastructure: The Role of Technologies Based Knowledge-intensive Business Services in National Innovation Systems*, SI4S project, European Commission (DG XII), TSER programme. Oslo: STEP Group.

Binswanger, H.P. (1978), *Induced Innovation*, Baltimore: Johns Hopkins University Press.

Bitner, M.-J., Booms, H.B. and Tetreault, M.S. (1990) 'The service encounter: diagnosing favorable and unfavorable incidents', *Journal of Marketing*, 54 (1): 71–84.

Björkegren, D. (1989) *Hur organisaationer lär*, Lund: Studentlitteratur.

Bogason, Peter (2001) *Fragmenteret forvaltning. Demokrati og netværksstyring i decentraliseret lokalstyre*, Aarhus: Systime.

Boje, D.A., Gephart Jr, R.P. and Thatchenkery, T.J. (eds) (1996) *Postmodern Management and Organization Theory*, Thousand Oaks: Sage.

Boland, R.J., Tenkasi, R.V. and Teeni, D. (1996) 'Designing information technology to support distributed cognition', in Meindl, J.R., Stubbart, C. and Porac, J.F. (eds) *Cognition Within and Between Organizations*, London: Sage.

Booz, Allen and Hamilton (1982) *New Product Management for the 1980s*, New York: Booz, Allen and Hamilton.

Borrás, S. (2000) *Science, Technology and Innovation in European Politics*, Roskilde: Roskilde University.

Borrás, S. and Johansen, H. (2001) 'The cohesion policy in the political economy of the EU, *Nordic Journal of International Studies*, 1: 39–6.

Boston, J. (1999) 'New models of public management: the New Zealand case', *Samfundsøkonomen*, (5): 5–13.

Bourdieu, P. and Wacqant, L. (1992) *An Invitation to Reflexive Sociology*, Chicago: University of Chicago Press.

Bowen, D. and Youngdahl, W. (1998) ' "Lean" service: in defence of a production-line approach', In *Marketing, Strategy, Economics, Operations and Human Resources: Insights on Service Activities. Proceedings from the 5th International Research Seminar in Service Management*. Aix-En-Provence: Institut d'administration des entreprises université d'Aix-Marsaille III.

Boyer, R. and Freyssenet, M., (forthcoming) *Le monde qui a changé la machine*, Oxford: Oxford University Press.

Brabander, B. de and Thiers, G. (1983) 'Een onderzoek naar de factoren die het succes van automatiseringsprojecten beïnvloeden', *Informatie*, 25 (12): 13–21.

Braverman, H. (1974) *Labor and Monopoly Capital*, New York: The Free Press.

Braczyk, H.-J., Cooke, P. *et al.* (eds) (1998) *Regional Innovation Systems*, London: UCL.

Brentani, U. de (1989) 'Success and failure in new industrial services', *Journal of Product Innovation Management*, 6 (6): 239–58.

Brentani, U. de (1991) 'Success factors in developing new business services', *European Journal of Marketing*, 25 (2): 33–59.

Brentani, U. (1993) 'The new product process in financial services: strategy for success', *International Journal of Bank Marketing*, 11 (3): 15–22.

Brown, R. (1991) 'Managing the "s" curves of innovation', *Journal of Marketing Management*, 7 (2): 189–202.

Brown, S., Bessant, J., Jones, P. and Lamming, R. (2000) *Strategic Operations Management*, Oxford: Butterworth-Heinemann.

Brown, S. and Eisenhardt, K.M. (1995) 'Product development: past research, present findings and future directions', *Academy of Management Review*, 20 (2): 343–78.

Bruner, J.S. (1965) *On Knowing: Essays for the Left Hand*, New York: Athenum.

Bruner, J.S. (1974) *Beyond the Information Given: Studies in the Psychology of Knowledge*, London: Allen and Unwin.

Burgelman, R.A. and Sayles, L.R. (1986) *Inside Corporate Innovation: Strategy, Structure and Managerial Skills*, New York: The Free Press.

Burns, T. and Stalker, G.M. (1961) *The Management of Innovation*, London: Social Science Paperbacks.

Cale, E.G. and Eriksen, S.E. (1994) 'Factors affecting the implementation outcomes of a mainframe software package: a longitudinal study', *Information and Management*, 26 (3): 165–75.

Callon, M. (1986) 'Eléments pour une sociologie de la traduction. La domestication des coquilles Saint-Jacques et des marins-pêcheurs dans la baie de Saint-Brieuc', *L'année sociologique*, 36: 169–208.

Callon, M. (1991) 'Réseaux technico-économiques et irréversibilité', in Boyer, R. (ed) *Les figures de l'irréversibilité en économie*, Paris: Edition de l'école des hautes études en science sociale.

Camp, R. (1989) *Benchmarking: The Search for Industry Best Practices That Lead to Superior Performance*, Milwaukee WI: Quality Press.

Campion, M. (1988) 'Interdisciplinary approaches to job design: a constructive replication with extensions', *Journal of Applied Psychology*, 73 (3): 467–81.

Caracostas, P. and Soete, L. (1997) 'The building of cross-border institutions in Europe: towards a European system of innovation?', in Edquist, C. (ed.) *Systems of Innovation: Technologies, Institutions and Organizations*, London: Pinter.

Carter, C. and Williams, B. (1957) *Industry and Technical Progress*, Oxford: Oxford University Press.

Chaffee, E. (1985) 'Three models of strategy', *Management Review*, 10 (1): 89–98.

Chao, G.T. and Kozlowski, S.W.J. (1986) 'Employee perceptions on the implementation of robotic manufacturing technology', *Journal of Applied Psychology*, 71 (1): 70–6.

Chaykowski, R.P. and Slotsve, G.A. (1992) 'The impact of plant modernization on organizational work practices', *Industrial relations*, 31 (2): 309–29.

Chesbrough, H.W. and Teece, D. (1996) 'When is virtual virtuous? Organizing for innovation', *Harvard Business Review*, 74 (1) January–February: 65ff.

Christensen, C.M. (1997) *The Innovator's Dilemma: When New Technologies Cause Great Firms to Fail*, Harvard, MA: Harvard Business School Press.

Clinch, P. (2000) 'Environmental policy reform in the EU', in Galli, G. and Pelkmans, J. (eds) *Regulatory Reform and Competitiveness in Europe*, Cheltenham: Edward Elgar.

Coch, L. and French, J.R.P. Jr (1948) 'Overcoming resistance to change', *Human Relations*, 1: 512–32.

Cohen, J. and Cohen, P. (1975) *Applied Multiple Regression/Correlation Analysis for the Behavioral Sciences*, Hillsdale NJ: Wiley.

Cohen, M.D. and March, J.G. (1974) *Leadership and Ambiguity: The American College President*, New York: McGraw-Hill.

Cohen, M.D., March, J.G. and Olsen, J.P. (1972) 'A garbage can model of organizational choice', *Administrative Science Quarterly*, 17 (1): 1–25.

Cohen, W.M. and Levinthal, D.A. (1989) 'Innovation and learning: the two faces of R-D', *Economic Journal*, 99 (397): 569–96.

Cohen, W.M. and Levinthal, D.A. (1990) 'A new perspective on learning and innovation', *Administrative Science Quarterly*, 35 (1): 128–52.

Cohen. W.M. and Levinthal, D.A. (1994) 'Fortune favours the prepared firm', *Management Science*, 40 (3): 227–51.

Cohendet, P., Kern, F., Mehmanpazir, B. and Munier, F. (1999) 'Knowledge coordination, competence creation and integrated networks in globalised firms', *Cambridge Journal of Economics*, 23 (2): 225–41.

Cohendet, P., Llerena, P., Stahn H. and Umbhauer G. (eds) (1998) *The Economics of Networks: Interaction and Behaviours*, Berlin: Springer.

Colombo, M., Garrone, P. (1996) 'Technological cooperative agreements and firm's R&D intensity: a note on causality relations', *Research Policy*, 25 (6): 923–32.

Commission of the European Communities (1989) *Business Services in the European Community: Situation and Role*, Bruxelles: Commission of the European Communities, III 89/2234/EN/Rev2.

Commission of the European Communities (1994) *Growth, Competitiveness, Employment. The Challenges and Ways forward into the 21st Century*, Luxembourg: Commission of the European Communities, Office for Official Publications.

Commission of the European Communities (1995) *Green Paper on Innovation*, Bruxelles: Commission of the European Communities.

Commission of the European Communities (1996) *The First Action Plan for Innovation in Europe*, Bruxelles: Office for Official Publications.

Commission of the European Communities (2000) *Towards a European Research Area*, Bruxelles: Commission of the European Communities.

Cooke, P., Boekholt, P. *et al.* (2000) *The Governance of Innovation in Europe: Regional Perspectives on Global Competitiveness*, London: Pinter.

Coombs, R., Richards A., Saviotti, P.P., and Walsh, V. (1996) *Technological Collaboration: The Dynamics of Cooperation in Industrial Innovation*, Cheltenham: Edward Elgar.

Cooper, R.G. (1986) *Winning at New Products*, Toronto: Holt.

Cooper, R. (1994) 'Third-generation new product processes', *Journa of Product Innovation Management*, 11 (1): 3–14.

Cooper, R. and Burrell, G. (1988), 'Modernism, postmodernism and organizational analysis: an introduction', *Organization Studies*, 9 (1): 91–112.

Cooper, R. and Kleinschmidt, E. (1986) 'An investigation into the new product process: steps, deficiencies, and impact', *Journal of Product Innovation Management*, 3 (2): 215–33.

Cooper, R. and Kleinschmidt, E.J. (1993) 'Major new products: what distingusihes the winners in the chemical industry?', *Journal of Product Innovation Management*, 10 (1): 90–111.

Cooper, R.B. and Zmud, R.W. (1990) 'Information technology implementation research: a technological diffusion approach', *Management Science*, 36: 123–39.

Coriat, B.and Weinstein, O. (1995) *Les nouvelles théories de l'entreprise*, Paris: Le livre de poche.

Cova, B. (1996) 'The postmodern explained to managers: implications for marketing', *Business Horizons*, November–December: 21–3.

Cram, L. (1997) *Policy-making in the EU*, London: Routledge.

Crawford, C.M. (1991) *New Product Management*, Homewood: Irwin.

Cronberg, T. (1990) *Fremtidsforsøg*, København: Akademisk Forlag.

Cronberg, T., Duelund, P., Jensen, O.M. and Qvortrup, L. (eds) (1991) *Danish Experiments: Social Constructions of Technology*, Copenhagen: New Social Science Monographs.

Crozier, M. (1964) *The Bureaucratic Phenomenon*, London: Tavistock.

Crozier, M. and Friedberg, E. (1977) *L'Acteur et le système. Les contraintes de l'action collective*, Paris: Éditions du Seuil.

Cullen, K. and Moran, R. (1992) *Technology and the Elderly: The Role of Technology in Prolonging the Independence of the Elderly in the Community Care Context*, Bruxelles: Commission of the European Communities: FAST research report: EUR 14419 EN.

Cummings, S. (1996) 'Back to the oracle: postmodern theory as a resurfacing of pre-modern wisdom', *Organization*, 3: 249–66.

Cyert, R.M. and March, J.G. (1963) *A Behavioral Theory of the Firm*, Englewood Cliffs: Prentice Hall.

Dalum, B., Johnson, B. *et al.* (1992) 'Public policy in the learning society', in Lundvall, B.-A. (ed.) *National Systems of Innovation: Towards a Theory of Innovation and Interactive Learning*, London: Pinter.

Daniel, W.W. (1987) *Workplace Industrial Relations and Technical Change Based on the DE/ESRC/PSI/ACAS Surveys*, London: Pinter.

Daniels, P. (1990) 'Geographical perspectives on the development of producer services', in Teare, R., Moutinho, L. and Morgan, N. (eds) *Managing and Marketing Services in the 90's*. London: Cassell.

Danish Ministy of Research (1994) *Info-samfundet år 2000. Rapport fra udvalget om Informationssamfundet år 2000*, Copenhagen: The Danish Ministry of Research.

Danish Ministy of Research (1996) *Info-samfundet for alle – den danske model*, Copenhagen: The Danish Ministry of Research

Davenport, T. (1993) *Process Innovation*, Boston: Harvard Business School Press.

David, P. (1985) 'Clio and the economics of QWERTY', *American Economic Review*, 75 (2): 332–7.

Davis, F.D. (1993) 'User acceptance of information technology: system characteristics, user perceptions and behavioral impacts', *International Journal of Man-Machine Studies*, 38 (3): 475–87.

Dean, J.W. Jr. and Snell, S.A. (1991) 'Integrated manufacturing and job design: moderating effects of organizational inertia', *Academy of Management Journal*, 34 (4): 776–804.

Decoster, E.and Matteaccioli, A. (1991) 'L'impact des réseaux d'innovation sur les milieux locaux: le rôle des réseaux des sociétés de conseil et des centres de recherche en Ile de France', *Revue d'économie régionale et urbaine*, (3/4): 479–508.

Delaunay, J.C.and Gadrey, J. (1992) *Services in Economic Thought: Three Centuries of Debate*, Dordrecht: Kluwer.

Dhebar, A. (1996) 'Speeding high-tech producer, meet the balking consumer', *Sloan Management*

Dijck, J.J.J. van (1974) *Organisatie in verandering. Sociologische modellen van veranderingsprocessen in organisaties (Changing Organizations)* Rotterdam: Universitaire Pers Rotterdam.

Djellal, F. (1995) *Changement technique et conseil en technologie de l'information*, Paris: L'Harmattan.

Djellal, F.and Gallouj, F. (1998) *Innovation in Services: The Results of a Postal Survey*. SI4S project, European Commission (DG XII), TSER program, Lille: IFRESI.

Djellal, F., Gallouj, F. and Gallouj, C. (1998) *Innovation Trajectories in French Service Industries*. SI4S project, European Commission (DG XII), TSER program, Lille: IFRESI.

Dodgson, M. (1993) *Technological Collaboration in Industry*, London: Routledge.

Dornblaser, B.M., Tse-Min Lin and Ven, A.H. van de (1989) 'Innovation outcomes: learning and action loops', in Van de Ven, A., Angle, H.L. and Scott Poole, M. (eds) *Research on the Management of Innovation: The Minnesota Studies*, New York: Harper and Row.

Dosi, G. (1982) 'Technological paradigms and technological trajectories: a suggested interpretation of the determinants and directions of technical change', *Research Policy*, 2 (3): 147–62.

Dosi G. (1988) 'Sources, procedures and microeconomic effects of innovation', *Journal of Economic Literature*, 25, September: 1120–71.

Dosi, G., Freeman, C., Nelson, R., Silverberg, G. and Soete, L. (eds) (1988) *Technical Change and Economic Theory*, London: Pinter.

Dougherty, D. (1996) 'Organizing for innovation', in Clegg, S.R., Hardy, C. and Nord, W.R. (eds) *Handbook of Organization Studies*, London: Sage.

Dougherty, D. and Hardy, C. (1996) 'Sustained product innovation in large, mature organizations: overcoming innovation-to-organization problems', *Academy of Management Journal*, 39 (5): 1120–53.

Drew, S. (1995) 'Strategic benchmarking: innovation practices in financial institutions', *International Journal of Bank Marketing*, 13 (1): 4–16.

Drucker, P. (1985) *Innovation and Entrepreneurship*, New York: Harper and Row.

Dutton, W. H., Blumler, J. G. and Kraemer, K. L. (eds) (1987) *Wired Cities: Shaping the Future of Communications*, Boston, MA: The Washington Program – Annenberg School of Communications, G.K Hall and Co.

Dyson, J. (1998) *Against the Odds*, London: Orion Books.

Easingwood, C. J. (1986) 'New product development for service companies', *Journal of Product Innovation Management*, 3, December: 264–75.

Edquist, C. (ed.) (1996) *Systems of Innovation: Technologies, Institutions and Organizations*, London: Pinter.

Edquist, C. and Hommen, L. (1998) *Government Technology Procurement and Innovation Theory*, Linköping: Linköping University – ISE Report.

Edvardsson, B. (1996) *Kvalitet och tjänsteutveckling*, Lund: Studentlitteratur.

Edvardsson, B. (1997) 'Quality in new service development: key concepts and a frame of reference', *International Journal of Production Economics*, 52: 31–46.

Edvardsson, B. and Gustafsson, A. (1999) 'Quality in the development of new products and services', in *The Nordic School of Quality Management*, Lund: Studentlitteratur.

Edvardsson, B. and Mattsson, J. (1992) 'Service design: a TQM instrument for service', Paper presented at the Service Productivity and Quality Challenge Conference, The Wharton School, University of Pennsylvania.

Edvardsson, B. and Thomasson, B. (eds) (1989) *Kvalitetsutveckling i privata och offentliga tjänsteföretag*, Stockholm: Natur och kultur.

Edvardsson, B., Haglund, L. and Mattsson, J. (1995) 'Analysis, planning, improvisation and control in the development of new services', *International Journal of Service Industry Management*, 6 (3): 24–35.

Egidi, M. (1997) 'Technological and organizational innovation as problem-solving activities', in Antonelli, G. and De Liso, N. (eds) *Economics of Structural and Technological Change*. London: Routledge.

Ehrenzweig, A. (1967) *The Hidden Order of Art: A Study in the Psychology of Artistic Imagination*, London: Weidenfeld and Nicolson

Eisenberg E.M. (1984) 'Ambiguity as strategy in organizational communication', *Communication Monographs*, 51: 227–42.

Elliot, K. and Roach, D. (1991) 'Are consumers evaluating your products the way you think and hope they are?', *Journal of Consumer Marketing*, 8 (1): 5–14.

Faulkner, W. and Senker, J. (1995) *Knowledge Frontiers*, Oxford: Clarendon.

Feenberg, A. (1992) 'From information to communication: the French experience with videotex', in Lea, M. (ed.) *Contexts of Computer-Mediated Communication*, Harvester Wheatsheaf.

Figuereido, P. (2000) *Teachnological Learning in Steel Plants in Brazil*, Brighton: SPRU, University of Sussex.

Fiol, M.C. (1996) 'Consensus, diversity and learning in organisations', in Meindl, J.R, Stubbart, C. and Porac, J.F. (eds) *Cognition Within and Between Organizations*, London: Sage.

Fitzgerald, B. (1998) 'An empirical investigation into the adoption of system development methodologies', *Information and Management*, 34 (6): 317–28.

Flanagan, J. C. (1954) 'The critical incident technique', *Psychological Bulletin*, 51 (4): 327–57.

Fleck, J. (1984) 'The employment effects of robots', in Lupton, T. (ed.) *Proceedings of the 1st International Conference on Human Factors in Manufacturing*, Kempston, England: IFS Publications and North-Holland.

Fleck, J. (1988) 'Innofusion or diffusation? The nature of technological development in robotics'. Edinburgh PICT Working Paper No. 7, Edinburgh: Edinburgh University.

Foray, D. (1993) *Modernisation des Entreprises, Coopération Industrielle Inter et Intra-Firmes et Ressources Humaines*. Report for the French Ministry of Research and Technology, Paris: Ministry of Research and Technology.

Foss N. J. (1998) 'The resource-based perspective: an assessment and diagnosis of problems', *Scandinavian Journal of Management*, 14 (3): 133–49.

Francis, D. (1994) *Step by Step Competitive Strategy*, London: Routledge.

Francis, D. (1997) *Partnerships with People*, London: Department of Trade and Industry.

Francis, J. (1994) 'Rethinking NPD; giving full rein to the innovator', *Marketing*, 26 (1): 6–7.

Fransman, M. (1994) 'Information, knowledge, vision and theories of the firm', *Industrial and Corporate Change*, 3 (3): 1–45.

Freeman, C. (1995) 'The national system of innovation in historical perspective', *Cambridge Journal of Economics*, 19 (1): 5–24.

Freeman C., Clark, J. and Soete, L. (1982) *Unemployment and Technical Innovation*, London: Pinter.

Freeman, C. and Soete, L. (1997) *The Economics of Industrial Innovation*, London: Pinter.

Freeman, R.E. (1984) *Strategic Management: A Stakeholder Approach*, Boston: Pitman.

French, W.L. and Bell, C.H. (1987) *Organization Development: Behavioural Science Intervention for Organization Improvement*, Englewood Cliffs: Prentice Hall.

Fuglsang, L. (2000) *Menneskelige ressourcer i hjemmehjælpen: fra pelsjæger til social entreprenør (Human resources in home-help: from trapper to social entrepreneur)*. Project 'Service development, internationalisation and competence development' (SIC). Case-report, Roskilde: Roskilde University.

Fuglsang, L. (2001) Management problems in welfare services: the role of the "social entrepreneur" in home-help for the elderly, the Valby case', *Scandinavian Journal of Management*, 17 (4): 437–55.

Gadrey, J. (ed). (1992) *Manager le conseil*, Paris: McGraw-Hill.

Gadrey, J. and Gallouj, F. (1994) *L'innovation dans l'assurance: le cas de l'UAP*. Report of the scientific department of l'UAP, Paris: Ministry of Research.

Gadrey, J.and Gallouj, F. (1998) 'The provider–customer interface in business and professional services', *Service Industries Journal*, 18 (2): 1–15.

Gadrey, J., Gallouj, F., Lhuillery, S., Ribault, T. and Weinstein, O. (1993) *La recherche-développement et l'innovation dans les activités de service: le cas du conseil, de l'assurance et des services d'information électronique*. Report for the French Ministry of

Higher Education and Research, Paris: French Ministry of Higher Education and Research.

Gallagher, M. and Austin, S. (1997) *Continuous Improvement Casebook*, London: Kogan Page.

Galli, G. and J. Pelkmans (eds) (2000) *Regulatory Reform and Competitiveness in Europe*, Cheltenham: Edward Elgar.

Gallouj, C. and Gallouj, F. (1996) *L'innovation dans les services*, Paris: Economica Poche.

Gallouj, F. (1994a) 'Cycles économiques et innovations de service: quelques inter-rogations à la lumière de la pensée schumpeterienne', *Revue française d'économie*, 9 (4): 169–213.

Gallouj, F. (1994b) *Economie de l'innovation dans les services*, Paris: L'Harmattan.

Gallouj, F.and Weinstein, O. (1997) 'Innovation in services', *Research Policy*, 26 (4–5): 537–56.

Garofoli G. (ed.) (1992) *Endogenous Development in Southern Europe*, Avebury: Aldershot.

Gattiker, U.E. and Larwood, L. (eds) (1995) *Technological Innovation and Human Resources: End-user Training*, New York: De Gruyter.

Garvin, D. (1993) 'Building a learning organisation', *Harvard Business Review*, (71) July–August: 78–91.

Gershuny, J. (1983) *Social Innovation and Division of Labour*, Oxford: Oxford University Press.

Gerwin, D. (1988) 'A theory of innovation processes for computer-aided manufac-turing technology', *IEEE Transactions on Engineering Management*, 35 (2): 90–100.

Geus, A. de (1996) *The Living Company* Boston: Harvard Business School Press.

Giddens, A. (1976) *New Rules of Sociological Method: A Positive Critique of Interpretative Sociologies*, London: Hutchinson and Co.

Giddens, A. (1984) *The Constitution of Society*, Cambridge: Polity.

Giddens, A. (1990) *The Consequences of Modernity*, Cambridge: Polity.

Giddens, A. (1991) *Modernity and Self-Identity*, Cambridge: Polity.

Gill, C. and Krieger, H. (1991) 'The diffusion of participation in new information technology in Europe: survey results', *Economic and Industrial Democracy*, 13: 331–58.

Glaser, B.G. and Strauss, A.L. (1967) *The Discovery of Grounded Theory: Strategies for Qualitative Research*. New York: Aldine.

Gordon, R. (1990) 'Système de production, réseaux industriels et régions: les trans-formations de l'organisation sociale et spatiale de l'innovation', *Revue d'économie industrielle*, 51 (1): 304–39.

Govarere, I. (1996) *The Use and Abuse of Intellectual Property Rights in EC Law*, London: Sweet and Maxwell.

Grabher, G. (1993) *The Embedded Firm: On the Socioeconomics of Industrial Networks*, London: Routledge.

Granovetter, M. (1973) 'The strength of weak ties', *American Journal of Sociology*, 78 (6): 1360–80.

Grant, R. (1991) 'The resource-based theory of competitive advantage: implications for strategy formulation', *California Management Review*, spring: 114 – 35.

Green, K. (1991) 'Shaping technologies and shaping markets: creating demand for biotechnology', *Technology Analysis and Strategic Management*, 3 (1): 57–76.

Greiner, L. and Metzger, R. 1983) *Consulting to Management*, New York: Heinemann.

Grindley, P., McBryde, R. and Roper, M. (1989) *Technology and the Competitive Edge: The Case of Richardson Sheffield*, London: London Business School.

Grönroos, C. (1990) *Service Management and Marketing: Managing the Moments of Truth in Service Competition*, New York: Lexington Books.

Guimaraes, T. and McKeen, J.D. (1993) 'User participation in information system development: moderation in all things', in Avison, D., Kendall, J.E. and de Gross, J.I. (eds) *Human, Organizational, and Social Dimensions of Information Systems Development. Proceedings of the IFIP wg8.2 Working Group Information Systems Development: Human, Social and Organizational Aspects*. Amsterdam: Elsevier Science Publishers B.V.

Gummesson, E. (1991) *Kvalitetsstyrning i tjänste- och serviceverksamheter – tolkning av fenomenet tjänstekvalitet och syntes av internationell forskning*. Research Report 91: 4, Karlstad: Service Research Center, University of Karlstad.

Gupta, A.K., Ray, S.P. and Wilemon, D.L. (1985) 'R&D and marketing dialogue in hi-tech firms', *Industrial Marketing Management*, 14 (1): 289–300.

Gustafsson, A., Ekdahl, F. and Edvardsson, B. (1999) 'Customer focused service development in practice: – a case study of Scandinavian Airlines System (SAS)', *International Journal of Service Industry Management*, 10 (4): 344–58.

Habermas, J. (1987) *The Theory of Communicative Action, Vol. Two: Lifeworld and System*, Cambridge: Polity.

Hage, J. (1980) *Theories of Organizations: Form, Process and Transformations*, New York: Wiley.

Hage, J. (1998) 'An overview of research', in Hage, J. (ed.) *Organizational Innovation*. Brookfield: Ashgate.

Hagedoorn, J. and Schakenraad, J. (1994) 'The effect of strategic technology alliances on company performance', *Strategic Management Journal*, 15 (4): 291–309.

Håkansson, H. (ed.) (1987) *Industrial Technology Development: A Network Approach*, London: Routledge.

Håkansson, H. (1989) *Corporate Technological Behaviour: Cooperation and Networks*, London: Routledge.

Hales, M. (1997) *Make or Buy in the Production of Innovation: Competences, Fullness of Services and the Architecture of Supply in Consultancy*. SI4S project, European Commission (DG XII), TSER program, Oslo: STEP Group.

Hamel, G and Prahalad, C.K. (1991) 'Corporate imagination and expeditionary marketing', *Harvard Business Review*, 70, July–August: 81–92.

Hamel, G. and Prahalad, C.K. (1994) 'Competing for the future', *Harvard Business Review*, 72, May–June: 122–8.

Hansen, P.A. and Serin, G. (1997) 'Will low technology products disappear?', *Technological Forecasting and Social Change*, (55): 179–91.

Hardy, C. and Clegg, S.R. (1996) 'Some dare call it power', in Clegg, S.R., Hardy, C. and Nord, W.R. (eds) *Handbook of Organization Studies*, London: Sage.

Hart, P. and Saunders, C. (1997) 'Power and trust: critical factors in the adoption and use of electronic data interchange', *Organization Science*, 8 (8): 23–42.

Hart, S. (1993) 'Dimensions of success in new product development: an exploratory investigation, *Journal of Marketing Management*, 9 (1): 23–41.

Hatch, M.J. (1998) *Organization Theory*, Oxford: Oxford University Press.

Hattrup, K. and Kozlowski, S.W.J. (1993) 'An across-organization analysis of the implementation of advanced manufacturing technologies', *Journal of High Technology Management Research*, 4 (2): 175–96.

Hauser, J. and Clausing, D. (1988) 'The house of quality', *Harvard Business Review*, (May–June): 63–73.

Hayes, R., S., Wheelwright, S. and Clark, K. (1988) *Dynamic Manufacturing: Creating the Learning Organisation*, New York: The Free Press.

Heap, J.P. (1989) *The Management of Innovation and Design*, London: Cassell.

Heath, C. (1997) 'The analysis of activities in face to face interaction using video', in Silverman, D. (ed.) *Qualitative Research: Theory, Method and Practice*. London: Sage.

Hedberg, B. (1981) 'How organizations learn and unlearn', in Nyström, P.C. and Starbuck, W.H. (eds) *Handbook on Organizational Design: Adapting Organizations to Their Environment*, Oxford: Oxford University Press.

Henderson, R.M. and Clark, K.B. (1990) 'Architectural innovation: the reconfiguration of existing product technologies and the failure of established firms', *Administrative Science Quarterly*, 35 (1): 9–30.

Henry, J. (ed.) (1991) *Creative Management*, London: Sage.

Henry, O. (1992) 'Entre savoir et pouvoir: les professionnels de l'expertise et du conseil', *Actes de la recherche en sciences sociales*, (95): 37–54.

Herbig, P., Howard, C. and Kramer, H. (1995) 'The installed base effect: implications for the management of innovation', *Journal of Marketing Management*, 11 (4): 387–401.

Herstatt, C. and Hippel, E. von (1992) 'Developing new product concepts via the lead user method', *Journal of Product Innovation Management*, 9 (3): 213–21.

Hines, P. (1994), *Creating World Class Suppliers: Unlocking Mutual Competitive Advantage*, London: Pitman.

Hippel, E. von (1976) 'The dominant role of users in the scientific instruments innovation process', *Research Policy*, 5 (3): 212–39.

Hippel, E. von (1988) *The Sources of Innovation*, Oxford: Oxford University Press.

Hippel, E. von and Thomke, S. (1999) 'Creating breakthroughs at 3M', *Harvard Business Review*, 77 (5): 47.

Hirschman, A. O. (1972) *Exit, Voice and Loyalty*, Cambridge MA: Harvard University Press.

Hirst, P. and Zeitlin, J. (1988) *Reversing Industrial Decline: Industrial Structure and Policy in Britain and Her Competitors*, Oxford: Berg.

Hirst, P. and Zeitlin, J. (1991) 'Flexible specialization versus post-Fordism: theory, evidence and policy implications', *Economy and Society*, 20 (1): 1–56.

Hodgson, G.M. (1998) *The Political Economy of Utopia: Why the Learning Economy Is Not the End of the History*, London: Routledge.

Hollinger, R. (1994) *Postmodernism and the Social Sciences: a Thematic Approach*, Thousand Oaks: Sage.

Holti, R., Neumann, J. *et al.* (1995) *Change Everything at Once: The Tavistock Institute's Guide to Developing Teamwork in Manufacturing*, London: Management Books 2000.

Horwitch, M. (ed.) (1986) *Technology in the Modern Corporation: A Strategic Perspective*, New York: Pergamon.

Houlder, V. (1994) 'Rewards for bright ideas', *Financial Times*, December: 18.

Humphrey, J. and Schmitz, H. (1996) 'The triple C approach to local industrial policy', *World Development*, 24 (12): 1859–77.

Illeris, S. (1996) *The Service Economy: A Geographical Approach*, West Sussex: Wiley.

Imai, M. (1997) *Gemba Kaizen*, New York: McGraw Hill.

Imai, K.-I., Nonaka, I. and Takeuchi, H. (1986) 'Managing the new product development process: companies learn and unlearn', in Clark, K.N. (ed.) *The Uneasy Alliance: Managing the Productivity–Technology Dilemma*, Boston: Harvard Business School Press.

Ives, B., Olson, M.H. and Baroudi, J.J. (1983) 'The measurement of user information satisfaction', *Communications of ACM*, 26(10): 785–93.

Isaksen, S.G. (ed.) (1987) *Frontiers of Creativity Research*, New York: Bearly.

Jæger, B. (1994) *Læring i organisationer – udvalgte teorier belyst med konkrete eksempler.* Copenhagen: AKF Forlaget.

Jæger, B. (1995) *Videotex i støbeskeen.* Tekster om teknologivurdering nr. 16, Copenhagen: Danish Technical University, Institut for Teknologi og Samfund.

Jæger, B. (2001) 'Some lessons of social experiments with technology', in Keeble, L. and Loader, B. (eds) *Community Informatics: Community Development Through the Use of ICT's*, London: Routledge.

Jæger, B. and Hansen, F.J.S. (1999) 'Multimedia in Denmark: an overview', in Williams, R. and Slack, R.S. (eds) *Europe Appropriates Multimedia: A Study of the National Uptake of Multimedia in Eight European Countries and Japan.* Norwegian University of Science and Technology. Report no. 42, 1999, Trondheim: Norwegian University of Science and Technology.

Jæger, B., Manniche, J. and Rieper, O. (1990) *Computere, lokalsamfund og virksomheder. Evaluering af Egvad Teknologiforsøg og Erhvervsprojektet i Ringkøbing Amt*, Copenhagen: AKF Forlaget.

Jæger, B. and Rieper, O. (1991) *Kommunal service som selvbetjening – evaluering af forsøg med videotex*, København: AKF Forlaget.

Jæger, B., Slack, R.S. and Williams, R. (2000) 'Europe experiments with multimedia: an overview of social experiments and trails', *The Information Society*, 16: 277–301.

Jaffe, A.B. (1986) 'Technological opportunity and spillovers of R-D: evidence from firms' patents, profits, and market value', *American Economic Review*, 76 (5): 984–1001.

Jewkes, J., Sawers, D. and Stillerman, R. (1969) *The Sources of Invention*, London: Macmillan.

Johne, A. and Harborne. P. (1985) 'How large commercial banks manage product innovation', *International Journal of Bank Marketing*, 3 (1): 54–70.

Johne, A. and Storey, C. (1998) 'New service development: a review of the literature and annotated bibliography', *European Journal of Marketing*, 32 (3/4): 184–251.

Johnson, B. (1992) 'Institutional learning', in Lundvall, B.-A.

Johnson, B. and Nielsen, K. (eds) (1998) *Institutions and Economic Change*, Cheltenham: Edward Elgar.

Johnson, M.D. (1998) *Customer Orientation and Market Action*, New Jersey: Prentice Hall.

Joshi, K. and Lauer, T.W. (1998) 'Impact of information technology on users' work environment: a case of computer aided design (CAD) system implementation', *Information and Management*, 34 (6): 349–60.

Kamien, M.I. and Schwartz, N.L. (1982) *Market Structure and Innovation*, Cambridge: Cambridge University Press.

Kano, N., Seraku, N., Takahashi, F. and Tsuji, S. (1984) 'Attractive quality and must-be quality', *Hinshitsu* 14 (2): 147–56 (in Japanese).

Kanter, R.M. (1983) *The Change Masters*, New York: Unwin.

Kanter, R.M. (1989) *When the Giants Learn to Dance*, London: Simon and Schuster.

Kardes, F.R. (1998) *Consumer Behaviour: Managerial Decision-making*, Reading, MA: Addison Wesley.

Katzenbach, J. and Smith, D. (1992) *The Wisdom of Teams*, Boston: Harvard Business School Press.

Kaufmann A. and Tödling. F. (2000) 'Systems of innovation in traditional industrial regions: the case of Styria in a comparative perspective', *Regional Studies*, 34 (1): 29–40.

Kharbanda, O. and Stallworthy, M. (1990), *Project Teams*, Manchester: NCC–Blackwell.

King, S. (1985) 'Has marketing failed or was it never really tried', *Journal of Marketing Management*, 1 (1): 1–19.

Kirzner, I. (1973) *Competition and Entrepreneurship*, Chicago: University of Chicago Press.

Klein, K.J. and Ralls, R.S. (1995) 'The organizational dynamics of computerized technology implementation: a review of the empirical literature, in Gattiker, U.E: and Larwood. L. (eds). *Technological Innovation and Human Resources: End-user Training*, New York: De Gruyter.

Klein, K.J. and Speer Sorra, J. (1996) 'The challenge of innovation implementation, *Academy of Management Review*, 21 (4): 1055–80.

Kleinknecht, A. and A. Reijnen, J.O.N. (1992) 'Why do firms cooperate on R-D?', *Research Policy*, 21 (4): 1–13.

Kline, S. (1985) 'Innovation is not a linear process', *Research Management*, 28 (4): 36–45.

Kline, S. and Rosenberg, N. (1986) 'An overview of innovation', in Landau, R. and Rosenberg, N. (eds) *The Positive Sum Strategy: Harnessing Technology for Economic Growth*, Washington DC: National Academy Press.

Knights, D. and Morgan, G. (1991) 'Strategic discourse and subjectivity: towards a critical analysis of corporate strategy in organizations', *Organization Studies*, 12 (2): 251–74.

Koestler, A. (1964) *The Act of Creation*, London: Hutchinson and Company.

Kolb, D. and Fry, R. (1975) 'Towards a theory of applied experiential learning', in Cooper, C. (ed.) *Theories of Group Processes*, Chichester: Wiley.

Korah, V. (1996) *Technology Transfer Agreements and the EC Competition Rules*, Oxford: Clarendon Press.

Kotler, P. (1983) *Principles of Marketing*, Englewood Cliffs: Prentice Hall.

Kotler, P. (1998) *Marketing Management*, Glencoe: Prentice Hall.

Kuczmarski, T.D. (1992) *Managing New Products: The Power of Innovation*, Englewoods Cliff: Prentice Hall

Kumar, K. (1995) *From Post-Industrial to Post-Modern Society*, Oxford: Blackwell.

Kumar, N., Scheer, L. and Kotler, P. (2000) 'From market driven to market driving', *European Management Journal*, 18 (2): 129–41.

Lai, V.S. and Mahapatra, R.K. (1997) 'Exploring the research in information technology implementation', *Information Management*, 32 (2): 187–201.

Lamming, R. (1993) *Beyond Partnership*, New York: Prentice Hall.

Landabaso, M. (1997) 'The promotion of innovation in regional policy: proposals for a regional innovation strategy', *Entrepreneurship and Regional Development*, 9: 1–24.

Langeard, E., Bateson, J.E., Lovelock, C.H. and Eiglier, P. (1981) *Services Marketing: New Insights from Consumers and Managers*. No. 81–104, Boston: Marketing Science Institute.

Lasch, S. (1990) *Sociology of Postmodernism*, London: Routledge.

Lave, J. and Wenger, E. (1991) *Situated Learning. Legitimate Peripheral Participation*, Cambridge MA: Cambridge University Press.

Lawrence, P. and Lorsch, J. (1967) *Organization and Environment: Managing Differentiation and Integration*, Boston: Harvard University Press.

Leonard, D. and Rayport, J.F. (1997) 'Spark innovation through empathic design', *Harvard Business Review*, November–December: 102–13.

Leonard-Barton, D. (1992) 'The organisation as learning laboratory', *Sloan Management Review*, 34 (1): 23–38.

Leonard-Barton, D. (1995) *Wellsprings of Knowledge: Building and Sustaining the Sources of Innovation*, Boston: Harvard Business School Press.

Levin, M. (1993) 'Technology transfer as a learning and development process: an analysis of Norwegian programmes on technology transfer', *Technovation* 13 (8): 497–518.

Levitt, B. and March, J. (1988) 'Organisational learning', *Annual Review of Sociology*, 14: 319–40.

Levitt, T. (1972) 'Production-line approach to service', *Harvard Business Review* 50 (5): 20–31.

Lieshout, M. van (2000) 'La ville virtuelle d'Amsterdam: entre domaine public et initiative privée', in Bastelaer, B. van, Henin, L. and Lobet-Maris, C. (eds).

Lieshout, M. van, Egyedi, T. M. and Bijker, W. E. (eds) (2001) *Social Learning Technologies: The Introduction of Multimedia in Education*, Aldershot: Ashgate.

Lincoln, J.W. (1962) 'Developing a creativeness in people', in Parnes S.J.and Harding H.F. (eds) *A Source Book for Creative Thinking*, New York: Charles's Scribners' Sons.

Lippman, S. and Rumelt, R.P. (1982) 'Uncertain imitability: an analysis of inter-firm differences in efficiency under competition', *Bell Journal of Economics*, 13, autumn: 418–38.

Lundvall, B.Å. (1985) *Production Innovation and User–Producer Interaction*, Aalborg: Aalborg University Press).

Lundvall, B.Å. (1988) 'Innovation as an interactive process: from user–producer interaction to the national system of innovation', in Dosi, G. *et al.*

Lundvall, B. Å. (1992a) (ed.) *National Systems of Innovation*, London: Pinter.

Lundvall B.Å. (1992b) 'User–producer relationships: national systems of innovation and internationalisation', in Lundvall, B.Å.

Lundvall, B. Å. (1998) 'The learning economy: challenges to economic theory and policies', in Johnson, B. and Nielsen, K. (eds) *Institutions and Economic Change*, Cheltenham: Edward Elgar.

Lundvall, B. Å. (1999) 'Knowledge production and the knowledge base', Aalborg: Dept of Business Studies, Aalborg Univerisity. Unpublished.

Lundvall B.Å. and Borras, S. (1997) *The Globalising Learning Economy: Implication for Innovation Policy*, EC: Luxembourg.

Luukkonen, T. (2000) 'Additionality of EU framework programmes', *Research Policy*, 29 (6): 711–724.

Lynn, G.S., Morone, J.G. and Paulson, A.S. (1997) 'Marketing and discontinuous innovation: the probe and learn process', in Tushman, M.L. and Anderson, P. (eds) *Managing Strategic Innovation and Change*, New York: Oxford University Press.

Lyotard, J.-F. (1986) *The Postmodern Condition: A Report on Knowledge*, Manchester: Manchester University Press.

Lyytinen, K. and Hirschheim, R. (1987) 'Information systems failure', *Oxford Surveys in Information Technology*, 4 (2): 257–309.

Machlup, F. (1980) *Knowledge: its Creation, Distribution and Economic Significance*, New York.

Madsen, O. Ø. (1986) 'Denmark', in Burns, P. and Dewhurst, J. (eds) *Small Business in Europe*, London: Macmillan.

Majchrzak, A. (1988) *The Human Side of Factory Automation*, San Francisco: Jossey-Bass.

Mangematin, V. and Nesta, L. (1999) 'What kind of knowledge can a firm absorb?', *International Journal of Technology Management*, 18 (3/4): 149–72.

March, J.G. and Olsen, J.P. (1989) *Rediscovering Institutions*, New York: The Free Press.

Mark-Herbert., D. and Nyström, H. (2000) *Technological and Market Innovation: A Case Study of the Development of a Functional Food-Proviva*. Report 133, Uppsala: Institute of Economics, SLU.

Martin, B. and Salter, A. (1996) *The Relationship between Publicly Funded Basic Research and Economic Performance*, Brighton: SPRU – University of Sussex.

Martin, C.R. and Horne, D.A. (1993) 'Service innovations: successful versus unsuccessful firms', *International Journal of Service Industry Management*, 4: 48–64.

Martin, C.R. and Horne, D.A. (1995) 'Level of success inputs for service innovations in the same firm', *International Journal of Service Industry Management*, 6 (4): 40–56.

Martin, J. (1995) 'Ignore your customer', *Fortune*, (8) May: 121–5.

Maskell, P. (1996) *Learning in the Village Economy of Denmark: The Role of Institutions and Policy in Sustaining Competitiveness*. DRUID Working Paper No. 96–6, Copenhagen.

Maskell P., Eskelinen, H., Hannibalsson, I., Malmberg, A. and Vatne, E. (1998) *Competitiveness. Localised Learning and Regional Development*, London: Routledge.

Mason, R.O. and Mitroff, I.I. (1981) *Challenging Strategic Planning Assumptions*, New York: Wiley.

Mattsson, J. (1992) *Företagsinterna utvecklingstjänster i verkstadsindustrin*. Research Report 92: 8, Karlstad: Service Research Center, University of Karlstad.

Mattsson, J. (1993) 'Service quality improvement: can local managers put in place their own systems', *New Zealand Journal of Business*, 15: 115–22.

Mattsson, J. (1994) 'Improving service quality in person to person encounters: a multi-disciplinary review', *The Service Industry Journal*, 14 (1): 45–61.

Mattsson, J. (2000) 'How to manage technology during services internationalisation', *The Service Industry Journal*, 20 (1): 22–39.

Mattsson, J. and Benson-Rea, M. (1993) 'Service quality improvement: can local managers put in place their own systems', *New Zealand Journal of Business*, 15 (1): 115–22.

Mazzoleni, R. and Nelson, R. (1998) 'The benefits and costs of strong patent protection: a contribution to the current debate', *Research Policy*, 27 (3): 273–84.

Meindl, J.R., Stubbart, C. and Porac, J.F. (eds). (1996) *Cognition Within and Between Organizations*, London: Sage.

Mensch, G. (1975) *Das technologische patt*, Frankfurt: Umschau.

Metcalfe, S (1998) *Evolutionary Economics and Creative Destruction*, London: Routledge.

Metcalfe, S. and Gibbons, M. (1989) 'Technology, variety and organisation: a systematic perspective on the competitive process', *Research on Technological Innovation, Management and Policy*, 4: 153–93.

Meeus, M.T.H. (1994) *Informatietechnologie: plaats en betekenis van arbeid. Theorie, empirie en praktijk, Structuratie van handelen in veranderingsprocessen* (*Information Technology: Role and Meaning of Working*) Tilburg: Tilburg University Press.

Meeus, M.T.H. and Oerlemans, L.A.G. (2000) 'Firm behaviour and innovative performance: an empirical exploration of the selection–adaptation debate', *Research Policy*, 29 (1): 41–58.

Miles, I., Kastrinos, N., Flanagan, K., Bilderbek, R., den Hertog, P., Huntink, W. and Bouman, M. (1994) *Knowledge-Intensive Business Services: Their Role as Users, Carriers and Sources of Innovation*, Manchester: PREST, University of Manchester.

Miller, D. and Droge, C. (1986) 'Psychological and traditional determinants of structure', *Administrative Science Quarterly*, 31 (2): 539–60.

Mintzberg, H. (1973) 'Strategy-making in three modes', *California Management Review*, 16 (2): 44–53.

Mintzberg, H. (1982) *The Structuring of Organizations*, Englewood Cliffs: Prentice Hall.

Mintzberg, H. (1989) *Mintzberg on Management*, New York: The Free Press.

Mintzberg, H. and Waters, J. (1982) 'Tracking strategy in an entrepreneurial firm', *Academy of Management Journal*, 25 (3): 465–99.

Mitchell, R. (1991) 'How 3M keeps the new products coming', in Henry, J. and Walker, D. (eds) *Managing Innovation*, London: Sage.

Montgomery, C.A. (1995) *Resource-based and Evolutionary Theories of the Firms*, London: Kluwer.

Moorman, C. and Slotegraaf, R.J. (1999) 'The contingency value of complementary capabilities in product development', *Journal of Marketing Research*, 36 (2): 239–49.

Morone, J. (1993) *Winning in High-tech Markets*. Cambridge, MA: Harvard Business School Press.

Mumford, E. (1986) 'Making technology work for people (vertaling M. Brummel), in Cornelis, P.A. and van Oorschot, J.M. (eds) *Automatisering met een menselijk gezicht*. Deventer: Kluwer.

Mumford, E. and Weir, M. (1979) *Computer Systems in Work Design: The ETHICS Method*, New York: Halsted (Wiley).

Myers, S. and Marquis, D.G. (1969) *Successful Industrial Innovation: A Study of Factors Underlying Innovation in Selected Firms*, Washington: National Science Foundation, NSF 69–117.

Nayak, P. and Ketteringham, J. (1986) *Breakthroughs: How Leadership and Drive Create Commercial Innovations That Sweep the World*, London: Mercury.

Nelson, R. (1993) *National Systems of Innovation*, Oxford: Oxford University Press.

Nelson, R. and Winter, S.G. (1982) *An Evolutionary Theory of Economic Change*, Cambridge MA: Belknap.

Ng, S.C.S., Pearson, A.W. and Ball, D.F. (1992) 'Strategies of biotechnology companies', *Technology Analysis and Strategic Management*, 4 (4): 351–61.

Nieminen, J. and Törnroos, J.Å. (1997) 'The role of learning in the evolution of business network in Estonia: four Finnish case studies', in Björkman, I. and Forsgren, M. (eds) *The Nature of the International Firm*, Copenhagen: Handelshøjskolens Forlag.

Nonaka, S. (1994) 'A dynamic theory of organizational knowledge creation', *Organization Science*, 5 (1): 14–37.

Nonaka, S. and Takeuchi, N. (1995) *The Knowledge-creating Company*, Oxford: Oxford University Press.

Norling, P., Edvardsson, B. and Gummesson, E. (1992) *Tjänsteutveckling och tjänstekonstruktion*. Research Report 92: 5, Karlstad: Service Research Center, University of Karlstad.

North, D. and Smallbone, D. (2000) 'The innovativeness and growth of rural SMEs during the 1990s', *Regional Studies*, 34 (2): 145–57.

Noteboom, B. (1992) 'Towards a dynamic theory of transactions', *Journal of Evolutionary Economics*, 2: 281–99.

Noteboom, B. (1999) 'Innovation, learning and industrial organisation', *Cambridge Journal of Economics*, 23 (2): 127–50.

Nutt, P.C. (1992) *Managing Planned Change*, New York: Macmillan.

Nyström, H. (1979) *Creativity and Innovation*, Chichester: Wiley.

Nyström, H. (1985) 'Product development strategy: an integration of technology and marketing', *Journal of Product Innovation Management*, 2 (1): 25–33.

Nyström, H. (1989) *Technological and Market Innovation: Strategies for Product and Company Development*, Chichester: Wiley.

Nyström, H. (1993) 'Creativity and entrepreneurship', *Creativity and Innovation Management*, 2 (4): 237–42.

Nyström, H. and Edvardsson, B. (1982) 'Product innovation in food processing: a Swedish survey', *R&D Management*, 12 (2): 67–72.

OECD (1995) *Trade in High Technology Industry*, Paris: OECD.

OECD (1996) *Employment and Growth in the Knowledge Based Economy*, Paris: OECD.

OECD (1996) *The Globalisation of Industry: Overview and Sector Report*, Paris: OECD.

OECD (1997) *Patents and Innovation in the International Context*, Paris: OECD.

OECD (1997a) *Globalisation and Small and Medium Enterprises (SMEs)*, vol. 2. Country Studies. Paris: OECD.

Oliver, N. (1996) *Benchmarking Product Development*, Cambridge: University of Cambridge.

Orlikowski, W.J. and Robey, D. (1991) 'Information technology and the structuring of organizations', *Information Systems Research*, 2 (1): 143–69.

Ortt, R.J. and Schoormans, P.L. (1993) 'Consumer research in the development process of a major innovation', *Journal of the Market Research Society*, 35 (4): 375–89.

O'Shea, A. and McBain, N. (1999) 'The process of innovation in small manufacturing firms', *International Journal of Technology Management*, 18 (5/6/7/8): 610–26.

Ovans, A. (1998) 'The customer doesn't always know best', *Market Research*, 7 (3): 12–14.

Pannenborg, A.E. (1986) 'Technology push versus market pull: the designers dilemma', in Roy, R. and Weild, D. (eds) *Product Design and Technological Innovation*, Milton Keynes: Open University Press.

Parasuraman, A., Zeithaml, V. and Berry, L. (1985) 'A conceptual model of service quality and its implications for future research', *Journal of Marketing*, 49 (4): 41–50.

Paulussen R.M.C., Wijers, G.M., van Delden, F., te Hennep, M., van der Velde, G.W.A., van Vooren, C. and Westeneng, J.P. (1993) 'Software-kwaliteit bespreekbaar maken' (Discussing software quality) Informatie, 34 (3): 127–138.

Pavitt, K. (1990) 'What we know about the strategic management of technology', *California Management Review*, 32: 17–26.

Pavitt, K. (1994) 'Key characteristics of large innovating firms', in Dodgson, M. and Rothwell, R. (eds) *The Handbook of Industrial Innovation*, Cheltenham: Edward Elgar, pp. 357–66.

Pavitt, K. (1998) 'The inevitable limits of EU R&D funding', *Research Policy*, 27 (6): 559–68.

Pedler, M., Boydell, T. *et al.* (1991) *The Learning Company: A Strategy for Sustainable Development*, Maidenhead: McGraw-Hill.

Penrose, E.T. (1959) *The Theory of the Growth of the Firm*, New York: Oxford University Press.

Perez, C. (1983) 'Structural change and the assimilation of new technologies in the economic and social system', *Futures*, 15 (4): 357–75.

Peterson, J. (1992) 'The European technology community: policy networks in a supranational setting', in Marsh, D. and Rhodes, R.A.W. (eds) *Policy Networks in British Government*, Oxford: Clarendon.

Pisano, G. (1994) 'Knowledge, integration and the locus of learning: an empirical analysis of process development', *Strategic Management Journal*, 15: 85ff.

Petit, P. (1986) *Slow Growth and the Service Economy*, London: Pinter.

Petit, P. (ed.) (1998) *L'économie de l'information: les enseignements des théories économiques*, Paris: La Découverte.

Pettigrew, A. (1985) *The Awakening Giant: Continuity and Change at ICI*, Oxford: Basil Blackwell.

Pettigrew, A. and Whipp, R. (1991) *Managing Change for Competitive Success*, Oxford: Blackwell.

Phillips, A. (1971) *Technology and Market Structure*, Lexington: Lexington Books.

Piaget, J. (1970) *Psychologie et épistémologie*, Paris: Denoël.

Pinchot, G. (1985) *Intrapreneuring*, New York: Harper and Row.

Pine, B. (1993) *Mass Customization*, Boston: Harvard Business School Press.

Piore, M.J. and Sabel, C.F. (1984) *The Second Industrial Divide*, New York: Basic Books.

Polanyi, H. (1958) *Personal Knowledge*, London.

Polanyi M. (1962) *The Tacit Dimension*, New York: Garden City.

Porter, M. (1980) *Competitive Strategy*, New York: The Free Press.

Porter, M. (1985) *Competitive Advantage*, New York: The Free Press.

Porter, M. (1990) *The Competitive Advantage of Nations*, New York: Macmillan.

Powell, W.W. (1990) 'Neither market nor hierarchy: network forms of organizations', *Research in Organizational Behavior*, 12: 295–336.

Powers, J.V. (1997) 'Scandinavian Airline Systems (SAS) measures customers' feedback through actions, words', *American Productivity & Quality Center*, June–July: 1–8.

Prahalad, C.K. and Hamel, G. (1990) 'The core competence of the corporation', *Harvard Business Review*, May–June: 79–91.

Pratten C. (1991) *The Competitiveness of Small Firms*, New York: Cambridge University Press.

Premkamur, G., Ramamurthy, K. and Nilakanta, S. (1994) 'Implementation of electronic data interchange: an innovation diffusion perspective', *Journal of Management Information Systems*, 11 (2): 157–86.

Preissl, B. (1995) 'Strategic use of communication technology: diffusion processes in networks and environments', *Information Economics and Policy*, 7: 75–100.

Quinn, J.B. (1985) 'Managing innovation: controlled chaos', *Harvard Business Review*, 63 (3): 73–84.

Quinn, J.B. (1992) *Intelligent Enterprise*, New York: The Free Press.

Quinn, J.B. (1999) 'Strategic outsourcing: leveraging knowledge capabilities', *Sloan Management Review*, 40 (4), summer: 9–21.

Rhodes, R.A.W. (1997) *Understanding Governance: Policy Networks, Governance, Reflexivity and Accountability*, Buckingham: Open University Press.

Rickards, T. (1999) *Creativity and the Management of Change*, Oxford: Blackwell.

Robey, D., Farrow, D. and Franz, C.R. (1989) 'Group process and conflict system development', *Management Science*, 35: 1172–91.

Robey, D., Smith, L.A. and Vijayasarathy, L.R. (1993) 'Perceptions of conflict and success in information systems development projects', *Journal of Management Information Systems*, 10 (1): 123–39.

Robinson, E.A. (1997) 'America's most admired companies', *Fortune*, (3) (March): 68–75.

Rogers, E. (1995) *Diffusion of Innovation*, New York: The Free Press.

Romm, T., Pliskin, N., Weber, Y. and Lee, A.S. (1991) 'Identifying cultural clash in MIS implementation, *Information and Management*, 24 (1): 99–109.

Rorty, R. (1980) *Philosophy and the Mirror of Nature*, Oxford: Blackwell.

Rosenberg, N. (1976) *Perspectives on Technology*, Cambridge: Cambridge University Press.

Rothwell, R. (1977) 'The characteristics of successful innovators and technically progressive firms (with comments on innovation research)', *Management Decision*, 7 (3): 191–206.

Rothwell, R. (1975) 'Intracorporate entrepreneurs', *Management Decision*, 13 (3): 35–43.

Rothwell, R. (1984) 'The role of small firms in the emergence of new technologies', in Freeman, C. (ed.) *Design, Innovation and Long Cycles in Economic Development*, London: Pinter.

Rothwell, R. (1992) 'Successful industrial innovation: critical success factors for the 1990s', *R&D Management*, 22 (3): 221–39.

Rothwell, R. and Bessant, J. (1992) *Fifth Generation Innovation and Fifth Wave Manufacturing*, Technology Transfer International, London: Teaching Company Directorate.

Rothwell, R. and Zegveld, W. (1982) *Innovation and Small and Medium-Sized Enterprises*, London: Pinter.

Rowley, T.J. (1997) 'Moving beyond dyadic ties: a network theory of stakeholder influences', *Academy of Management Review*, 22 (4): 887–910.

Rumelt, R.P. (1987) 'Theory, strategy, and entrepreneurship', in Teece, D.J. (ed.) *The Competitive Challenge: Strategies for Industrial Innovation and Renewal*, Cambridge MA: Harvard Business School Press.

Rush, H., Brady, T. and Hobday, M. (1997) *Learning between Projects in Complex Systems*. Working Paper, Brighton: Centre for the Study of Complex Systems, University of Brighton.

Sabel, C. and Zeitlin, J. (1997) *World of Possibilities*, New York: Cambridge University Press.

Salançik, G.R. (1977) 'Commitment is too easy', *Organizational Dynamics*, 6 (1): 62–80.

Sanz, L. and Borrás, S. (2000) 'Explaining changes and continuity in the EU technology policy: the politics of economic ideas', in Dresner, S. and Gilbert, N. (eds) *The Dynamics of European Science and Technology Policies*, Aldershot: Ashgate.

Saviotti, P.P. (1996) *Technological Evolution, Variety and the Economy*, Cheltenham: Edward Elgar.

Schein, E. (1984) 'Coming to a new awareness of organisational culture', *Sloan Management Review*, winter: 3–16.

Schein, E. (1985) *Organizational Culture and Leadership*, San Francisco: Jossey-Bass.

Scheuing, E. and Johnson, E. (1989) 'A proposed model for new service development', *The Journal of Service Marketing*, 3 (2): 25–34.

Schmitz, H. (1999) 'Collective efficiency and increasing returns', *Cambridge Journal of Economics*, 23 (4): 465–83.

Schumpeter, J. (1934) *The Theory of Economic Development*, Harvard: Oxford University Press.

Schumpeter, J. (1939) *Business Cycles*, New York: McGraw-Hill.

Schumpeter, J. (1943) *Capitalism, Socialism and Democracy*, London: Unwin.

Schutz, A. (1967) *The Phenomenology of the Social World*, Evanstone IL: Northwestern University Press.

Scott, G. (1997) 'Technology notes. Management of technology. Delphi study: a summary', *Technology Studies*, 4 (2): 165–6.

Senge, P. (1990) *The Fifth Discipline: The Art and Practice of the Learning Organization*, New York: Century Business.

Sengenberger, W. (1989) 'Small firms development and industrial organisation', in Poutsma, E. and Walravens, A. (eds) *Technology and Small Enterprises*, Stevinweg: Delft University Press.

Sexton, D.L. and Kasarda, J. (eds) (1992) *The State of the Art of Entrepreneurship*, Boston MA: PWS–Kent

Sharp, M. (1991) 'The single market and European technology policies', in Freeman, C., Sharp, M. and Walker, W. (eds). *Technology and the Future of Europe: Global Competition and the Environment in the 1990s*, London: Pinter.

Sharp, M. (1998) 'Competitiveness and cohesion: are the two compatible?', *Research Policy*, 27 (6): 569–88.

Shillito, M. (1994) *Advanced QFD: Linking Technology to Market and Company Needs*, New York: Wiley.

Shostack, G.L. (1984) 'Service design in the operating environment', in George, W.R. and Marshall, C.E. (eds) *Developing New Services*, Chicago: American Marketing Association.

SIC (1999) *Danish Service Firms' Innovation Activities and use of ICT Based on a Survey*, Report no. 2 from the Project 'Service Development, Internationalisation and Competence Development', Roskilde: Centre of Service Studies, Roskilde University.

Silverstone, R. and Hirsch, E. (eds) (1992) *Consuming Technologies: Media and Information in Domestic Spaces*, London: Routledge.

Simon, H.A. (1989) *Scientific Discovery as Problem Solving*, New York: Peano Lecture.

Slaughter, S. (1998) 'Innovation and learning during implementation: a comparison of user and manufacturer innovations', *Research Policy*, 22 (1): 81–95.

Smith, M.R. and Marx. L. (eds) (1994) *Does Technology Drive History? The Dilemma of Technological Determinism*, Cambridge MA: MIT Press.

Smith, P. and Reinertsen, D. (1991) *Developing Products in Half the Time*, New York: Van Nostrand Reinhold.

Stacey, R.D. (1993) *The Chaos Frontier: Creative Strategic Control for Business*, Oxford.

Stauss, B. and Hentschel, B. (1991) 'Attribute-based versus incident-based measurement of service quality: results from an empirical study in the German car service industry', paper presented at the EIASM workshop on Service Quality, Brussels, 16–17 May.

Staten, H. (1984) *Wittgenstein and Derrida*, London: University of Nebraska Press.

Steen, M. v. d. (2000) 'Technology policy learning: towards endogenous government in the analysis of systems of innovation'. Unpublished paper.

Sterlacchini, A. (1999) 'Do innovative activities matter to small firms in non-R&D-intensive industries? An application to export performance', *Research Policy*, 28 (8): 819–32.

Sternberg, R.J. (ed.) (1988) *The Nature of Creativity*, New York: Cambridge University Press.

Stone, E.F. and Hollenbeck, J.R. (1989) 'Clarifying some controversial issues surrounding statistical procedures for detecting moderator variables: empirical evidence and related matters', *Journal of Applied Psychology*, 74 (1): 3–10.

Storper, M. (1996) 'Institutions of the knowledge-based economy', in OECD, *Employment and Growth in the Knowledge-based Economy*, Paris: OECD.

Strange, S. (1998) 'Who are EU? Ambiguities in the concept of competitiveness', *Journal of Common Market Studies*, 36 (1): 101–14.

Sundbo, J. (1992) 'The tied entrepreneur', *Creativity and Innovation Management*, 1 (3): 109–20.

Sundbo, J. (1994) 'Modulization of service production', *Scandinavian Journal of Management*, 10 (3): 245–66.

Sundbo, J. (1996) 'Balancing empowerment', *Technovation*, 16 (8): 397–409.

Sundbo, J. (1998a) *The Organisation of Innovation in Services*, Copenhagen: Roskilde University Press.

Sundbo, J. (1998b) *The Theory of Innovation*, Cheltenham: Edward Elgar.

Sundbo, J. (1999) 'Empowerment of employeees in small and medium sized service firms', *Employee Relations*, 21 (2): 105–27.

Sundbo J. and Fuglsang, L. (2002) 'Innovation as strategic reflexivity', in Sundbo, J. and Fuglsang, L. (eds) *Innovation as Strategic Reflexivity*, London: Routledge.

Sundbo, J. and Gallouj, F. (1999) *Innovation in Seven European Countries*. Report 99: 1, Roskilde: Centre of Service Studies, Roskilde University.

Sundbo, J. and Gallouj, F. (2000) 'Innovation as a loosely coupled system in services', *International Journal of Services Technology and Management*, 1 (1): 15–36.

Szebehely, M. (1995) *Vardagens organisering. Om vårdbiträden och gamla i hemtjänsten*, Lund: Studentlitteratur.

Tarde, G. de (1993) *Les lois de l'imitation*, Paris: Kimé (first edition 1890).

Tauber (1974) 'Predictive validity in consumer research', *Journal of Advertising Research*, 15 (5): 59–64.

Teece, D. and Pisano, G. (1994) 'The dynamic capabilities of firms: an introduction', *Industrial and Corporate Change*, 3 (3): 537–55.

Terril, C.A. (1992) 'The Ten Commandments of new service development', *Management Review*, 81 (2): 24–7.

Teubal, M. (1976) 'On user needs and need determination: aspects of the theory of technological innovation', *Research Policy*, 5 (2): 266–89.

Thamhain, H. and D. Wilemon (1987) 'Building high performance engineering project teams', *IEEE Transactions on Engineering Management*, 34 (3): 130–7.

Thompson, J.D. (1956) 'Authority and power in identical organizations', *American Journal of Sociology*, 62 (2): 290–301.

Thompson, J. D. (1976) *Organizations in Action*, New York: McGraw Hill.

Thrift, J. (1997) 'Too much good advice', *Marketing*, 3 (3): 18.

Thwaites, D. (1992) 'Organizational influences on the new product development process in financial services', *Journal of Product Innovation Management*, 9 (4): 303–13.

Tidd, J., Bessant, J. and Pavitt, K. (1997) *Managing Innovation*, Chichester: Wiley.

Tijssen, R.J.W. and E. van Wijk (1999) 'In search of the European paradox: an international comparison of Europe's scientific performance and knowledge flows in information and communication technologies research', *Research Policy*, 28 (5): 519–43.

Trott, P. (1998a) 'Growing businesses by generating genuine business opportunities: a review of recent thinking', *Journal of Applied Management Studies*, 7 (2): 211–22.

Trott, P. (1998b) *Innovation Management and New Product Development*, London: Prentice Hall.

Trott, P. *et al.* (1995) 'Inward technology transfer as an interactive process: a case study of ICI, *Technovation*, 15 (1): 25–43.

Tushman, M. and Anderson, P. (1987) 'Technological discontinuities and organizational environments' *Administrative Science Quarterly*, 31 (3): 439–65.

Tyre, M.J. and Orlikowski, W.J. (1997) 'Exploiting opportunities for technological improvement in organizations', in Tushman, M. and Anderson, P. (eds) *Managing Strategic Innovation and Change*, New York: Oxford University Press.

UNIDO (1995) *Technology Transfer Management*, Vienna: United Nations Industrial Development Organisation.

Urban, G.L. and Hauser, J.R. (1980) *Design and Marketing of New Products*, Englewood Cliffs: Prentice Hall.

Utterback, J. (1994) *Mastering the Dynamics of Iinnovation*, Boston: Harvard Business School Press.

Väyrynen, R. (1998) 'Global interdependence or the European fortress? Technology policies in perspective', *Research Policy*, 27 (6): 627–32.

Ven, A.van den, Angle, H. and Poole, H. (1989) *Research on the Management of Innovation*, New York: Harper and Row

Ven, A.H. van den and Rogers, E.M. (1988) 'Innovations and organizations: critical perspectives', *Communication Research*, 15 (5): 632–51.

Vestey, R. (2000) 'The role of IT analysts in the IT industry'. Dissertation, Portsmouth: Business School, University of Portsmouth.

Volberda, H. (1998) *Building the Flexible Firm*, Oxford: Oxford University Press.

Wall, T.D., Corbett, J.M., Martin, R., Clegg, C.W. and Jackson, P.R. (1990) 'Advanced manufacturing technology, work design, and performance: a change study', *Journal of Applied Psychology*, 75 (6): 691–7.

Wallas, G. (1926) *The Art Of Thought*, New York: Harcourt Brace.

Warmerdam, J. and Riesewijk, B. (1988) *Het slagen en falen van automatiseringsprojecten, Een onderzoek naar de sociaal-organisatorische implicaties van automatisering voor gebruiker-sorganisaties en computer service bedrijven*, Nijmegen: ITS.

Warnecke, H.-J. (1992) *The Fractal Company*, Berlin: Springer-Verlag.

Wentink, T. and Zanders, H. (1985) *Kantoren in actie. Een onderzoek naar kantoorautoma-tisering en de gevolgen voor kantoorarbeid en kantoororganisatie*, Deventer: Kluwer.

Wheelwright, S.E. and Clark, K.B. (1999) *Revolutionizing Product Development*, New York: The Free Press.

Whiston, T. (1996) 'Knowledge and sustainable development: towards the furtherance of a global communication system', in Gill, K. (ed.) *Information Society*, London: Springer.

Wijnen, B.J. and Oostrum, J. van (1993) 'Een contingentiemodel voor de invoering van geautomatiseerde informatiesystemen', *Mand O Tijdschrift voor organisatiekunde en sociaal beleid*, 47 (2): 88–103.

Wilhelmsson, M. and Edvardsson, B. (1994) *Service Development (In Swedish)*, Karlstad: Service Research Center, University of Karlstad.

Willener, A. (1967) *Interprétation de l'organisation dans l'industrie. Essai de sociologie de changement*, Paris: Mouton.

Williams, R. and Edge, D. (1996) 'The social shaping of technology', *Research Policy*, 25 (6): 865–99.

Williams, R., Slack, R. and Stewart, J. (2000) *Social Learning in Multimedia: Final Report*. EC Targeted Socio-Economic Research. Project: 4141 PL 951003, Edinburgh: University of Edinburgh, Research Center for Social Sciences.

Witkin, H.A., Dyk, R.B., Faterstone, H.F., Goodenough, S.A. and Karp, S.A. (1962) *Psychological Differentiation*, New York: Wiley.

Womack, J., Jones, D. and Roos, D. (1991) *The Machine That Changed the World*, New York: Rawson Associates.

Woolgar, S. (1991) 'Configuring the user: the case of usability trials', in Law, J. (ed.) *A Sociology of Monsters: Essays on Power, Technology and Domination*, London, Routledge.

Zahra, S.A., Nash, S. and Bickford, D.J. (1995) 'Transforming technological pioneering into competitive advantage', *Academy of Management Executive*, 9 (1): 17–32.

Zammuto, R.F. and O'Connor, E.J. (1992) 'Gaining advanced manufacturing technologies' benefits: the roles of organization design and culture', *Academy of Management Review*, 17 (4): 701–28.

Zeithaml, V., Berry, L. and Parasuraman, A. (1985) 'Problems and strategies in service marketing', *Journal of Marketing*, 49 (1): 34–46.

Zucker, (1996) 'The institutionalization of institutional theory', in Clegg, S.R., Hardy, C. and Nord, W.R. (eds) *Handbook of Organization Studies*, London: Sage.

Index

(Authors are indexed only if they are quoted for specific concepts, ideas or sentences. Contributors in this book are not indexed.)

Lightning Source UK Ltd.
Milton Keynes UK
UKOW03n0624151213

223008UK00006B/150/P